Daytime Star

By the same author

EXPLORING THE GALAXIES
THE CRAB NEBULA

Daytime Star

The Story of Our Sun

SIMON MITTON

Charles Scribner's Sons/New York

First Charles Scribner's Sons Paperback Edition 1983

Copyright © 1981 Simon Mitton

Library of Congress Cataloging in Publication Data

Mitton, Simon, 1946–
  Daytime star, the story of our sun.

  Bibliography: p.
  Includes index.
  1. Sun—Popular works. I. Title.
QB521.4.M57   523.7   81–669
ISBN 0-684-17829-X   AACR2

1 3 5 7 9 11 13 15 17 19 F/P 20 18 16 14 12 10 8 6 4 2

Printed in the United States of America

Contents

Illustrations

IMPORTANT WARNING

This is a book about the Sun. Under no circumstances must anyone attempt to look at the Sun directly, even through filters or smoked glass. The Sun is an intense source of ultraviolet and infrared radiation which, if concentrated on the human eye, may cause irreversible damage. The book describes, on page 33, a safe method of observing the Sun by projecting an image onto a sheet of white card.

Introduction

Among the billions of stars in the Milky Way the Sun is unremarkable. When we consider the incomprehensible vastness of the cosmos, with its numerous clusters of galaxies stretching on and on through space, we are struck with the insignificance of our planet, our Sun, even our Galaxy, when set against the silent backdrop of the Universe. We see the giant disc of our Galaxy as a veil of light draped across the sky. In reality it is a flat disc of stars, arranged with two spiral arms emerging from the centre. Our Sun burns far out along one of these spiral arms. It is some 32,000 light years from the heart of the Milky Way. Within the Galaxy there must be billions of stars like it, for it has an average size, typical mass, normal structure, temperature and luminosity relative to stars in general. The only reason astronomers have for taking an interest in it is the simple one that we are orbiting this particular star and can therefore study it in great detail. No other star is close enough for us to see fine structure on its visible surface. No other star definitely has several planets, although this seems likely on theoretical grounds. It is wrong to think that all astronomers work in the dark; in fact about one-fifth of them are engaged in research on the daytime star.

Modern research on the Sun involves many scientific disciplines apart from traditional astronomy. Solar energy is the driving force behind natural phenomena on the Earth: it heats the atmosphere, oceans and landmasses. As the engine that drives the climate it must be connected in some way with climatic change—droughts, ice ages, and so forth. Particles blasted away from the Sun in energetic outbursts, such as solar flares, are responsible for the auroral displays studied by atmospheric scientists. The solar interior is a unique laboratory for physicists. Within it there are regimes of temperature, pressure and density such as could never be created on the Earth. By peering at the Sun we can collect information on the fluid motions of hot gases entangled in magnetic fields. It was by looking at the Sun that we learned how nuclear energy is released deep in the interiors of stars. The Sun also holds secrets on the chemical composition of the universe. It condensed from a cosmic gas cloud. We can deduce the chemical make-up of that birthplace by looking at the elements present in the Sun today.

Solar research is a vigorous branch of astronomy. Throughout the 1960s and 1970s it tended to be overshadowed by the exotic array of objects discovered by radio astronomers and X-ray astronomers. Radio galaxies, quasars, pulsars, neutron stars and black holes entered the ordinary language as scientists found phenomena that seemed time after time to conflict with established theories. While all this was going on, solar observatories quietly continued their work. Then, on 14 May 1973, NASA launched Skylab from Florida. This orbiting laboratory dedicated to the study of the Sun revolution ized modern solar physics. Its data base will be worked on for years to come. A cluster of six on-board telescopes scrutinized the Sun across a broad range of wavelengths, including the X-rays and ultraviolet light that is blocked by our atmosphere. In all the station was manned for 171 days. It exceeded all the targets set for operation, and was a major scientific achievement of the 1970s.

It was also during the 1970s that concern about sources of energy became sharply focussed. There is no energy shortage as such, merely a shortage of cheap and readily-available energy. The Sun provides enormous amounts of thermal energy free-of-charge and without the side effects associated with the burning of fossil fuels or the radiation hazards of nuclear-energy release. Any rational energy policy will in the long run have to include much greater use of solar energy. This is now generally accepted, and that acceptance has served to increase interest in solar studies in general.

In this book I have attempted to outline our present understanding of the Sun. This is a fascinating story, involving history, physics, astrophysics, space research and energy research. The unifying theme is that of physics as a discipline that accounts for many of the observed phenomena. At the same time I have indicated how at certain points in the development of physical theory observations of the Sun proved to be of crucial importance.

In general I have not given acknowledgement in the main text to all the individual researchers whose work I have consulted. The exceptions are those individuals who seem to me to have made particularly novel contributions. Those readers who wish to explore further the world of the Sun may find that the short bibliography at the end of the book gives a starting point to the extensive literature.

An Ancient Sun

Throughout ancient times the peoples of the Earth worshipped the Sun, almost the only source of heat and light. Even today this tradition probably continues among a few primitive tribes still left relatively undisturbed in remote desert and jungle communities. When we think of the building projects undertaken thousands of years ago we are reminded of the mysterious structures that have survived to the present day: the great pyramids in Egypt, the lonely stone circles of the megalithic age in western Europe, and the ruins of the Mayan civilization, for example. Why did those communities put so much effort into the construction of enduring monuments? We don't really know the answer, but certainly the need to carry out sun-watching played a major part in the design of the monuments.

Today's counterparts of those ancient temples are the great solar observatories. Our Sun sustains every living creature and plant on our home planet. It influences the climate and poses serious dangers to humans working in space. And so the Sun is kept under constant surveillance by astronomers. They want to understand more about how it works and just what effects it has on us and our environment. In order to gather this information about the Sun scientists operate several special observatories which are sited in locations that have sparkling clear days for most of the year. Instruments measure the magnetic field and radiation from the Sun, often on a continuous basis.

Astronomers have found that our Universe is of immense size and great age, seemingly boundless and timeless to the mind of man. Out and out it reaches, galaxies separated by millions of light years, whole galaxy clusters stranded in the empty blackness of the cosmos. Our Milky Way is just one of these galaxies, seen from within as a blur of stardust splashed across the dome of the sky. Yet here, our local neighbourhood in space, there are a hundred billion stars. Arranged in a spiral galaxy, this pinwheel of suns is a hundred thousand light years from side to side. In one of these spiral arms is our Sun, located

The McMath solar tower at the Kitt Peak National Observatory in Arizona, one of the major facilities available to the modern solar astronomer wishing to follow the behaviour of the Sun. (Kitt Peak National Observatory, Arizona)

far from the limb of the Milky Way. It has only one truly unique feature: it is *our* Sun, *our* daytime star. It is nearer than any other star, yet a passenger airliner would take twenty years to reach it, were such a journey possible at all. This distance, so vast for us, is covered in a little over eight minutes by the heat and light rays leaving its surface and warming our planet.

In this book I have set out the story of the Sun as it is understood by modern astronomy. Our awareness of the Sun's importance has been increased by the awesome predictions that Earth's reserves of fossil fuel (oil and coal) will be exhausted in the foreseeable future. The only long-term solution to this inescapable problem is to make intelligent (and safe) use of both nuclear power and solar energy. Our own future will be determined by the success of scientists in harnessing the Sun's enormous energy emission. This growing awareness of the Sun's potential has led to an increase in research activity and to dramatic new findings about the nature of the Sun. We can now 'see' the Sun across a very broad window of the spectrum, taking in X-rays at one extreme and radio waves at the other. All this has caused a revolution in solar studies. But let's begin our survey of the Sun back in ancient times, when, as we shall see, many of the achievements were at least as impressive as those of the present day.

Five thousand years ago in southern England the Neolithic culture started serious astronomical work. Its forefathers had crossed what is today the English Channel by means of a landbridge connecting the British Isles to mainland Europe. In about 10,000 BC the rising waters of the North Sea cascaded across the bridge and stranded the hunters of the Middle Stone Age in the British Isles. Gradually a transition from a hunter-gatherer to a farming society took place. By 3000 BC the Neolithic (New Stone Age) society had emerged, and with it industrious cultures that cleared the forest and scrub lands, planted and harvested grain, and developed a technology to move giant blocks of stone. The development of agriculture led to more settled communities in which there must have been more incentive for co-operation and more time for thought.

In about 2600 BC, we can't be sure of the precise date, the Neolithic people of southern England dug out a circular ditch on the gently undulating plain near to the city of Salisbury. The ditch is 105 metres (350 feet) across and exactly circular. All the rubble from the ditch they heaped into two banks on each side of the ditch. Nearly 3,000 cubic metres of crumbly chalk had to be hacked out with picks and shovels made from animal bones. Actually, the shoulder blade of an ox makes a rather good shovel. Even so, around ten man-years of labour went into throwing up the two chalk ramparts. At the northeast point of the circle they left an entrance, and 30 metres (100 feet) beyond this gap they erected an upright stone now called the Heel Stone. This famous block of sandstone is 6 metres from end to end and weighs 35 tons. However, moving it 32 kilometres (20 miles) from the Marlborough Downs, which were then strewn with large boulders, to its present location would not have seemed enormously difficult to the resourceful builders of the New Stone Age. They got the stone into a hole in the chalk, but it has heeled over by 30 degrees in the last five millennia.

Inside the circular banks two mounds were made, and four stones erected at the corners of a rectangle. Finally they scooped out 56

STONEHENGE

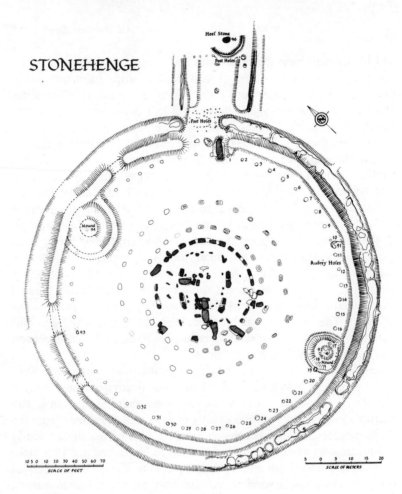

Ground plan of Stonehenge showing the outer ditch and bank which surrounds the impressive stone monument. The Heel Stone, coded 96 here, was used to mark the midsummer sunrise. Four markers at positions 91, 92, 93 and 94 indicated the crucial astronomical rectangle whose sides furnished important sightlines for sunrise, and sunset, as well as critical lunar sightlines. The 56 Aubrey holes, the first 32 of which are numbered on the plan, may have been a counting device for keeping track of eclipses. Only at the latitude of Stonehenge can such a symmetrical monument be laid out. (Her Majesty's Stationery Office, Crown Copyright)

holes, exactly dividing a perfect circle into 56 equal segments. Not bad work for illiterate people doing practical geometry 2,000 years before Euclid, the founder of geometry as an exact science. But what was it all for?

From within, the circular walls would have made a blinding white barrier, enclosing the sacred area, with its 56 holes and precise rectangles. The rest of Stonehenge, the giant archways and megaliths that strike the visitor today, were not to be erected for a further thousand years.

Gerald Hawkins cracked the mystery of Stonehenge in 1963. In an iconoclastic report published by the austere scientific weekly *Nature*, he set out a truly daring and controversial theory: that Stonehenge was constructed as an astronomical observatory for keeping track of the Sun and the Moon. This endeavour apparently had two purposes: to regulate the calendar, and thus keep agricultural affairs in good order; and to predict future eclipses of the Sun.

To explain these theories of Stonehenge, which are applicable to other standing-stone circles, I must first describe the motion of the Sun and Moon, as perceived by the Stonehenge builders. The Sun rises at different points on the horizon on different days of the year. In the northern hemisphere it rises at its most northerly point on the horizon on midsummer day and at the most southerly point on midwinter day. This immediately suggests a method of checking the calendar. By carefully noting the rising of the Sun on successive mornings they could identify the turning points of the yearly cycle by suitable markers. Two wooden sticks positioned tens of metres apart could act as sightlines, like a gunsight, to the horizon. After a few years of practice the critical sunrise (and sunset) lines could be permanently recorded by the positions of stout posts or even stone markers. The first builders at Stonehenge apparently did just this, for an observer in the centre of the rectangle sees the midsummer sunrise take place over the top of the Heel Stone. Two sides of the rectangle also furnish these same sightlines.

Further alignments present in the geometry of Stonehenge are even more impressive. Looking in the opposite direction, back along the summer sunrise lines, takes the eye to the point on the south horizon where the Sun *sets* on midwinter days.

The most impressive of all natural phenomena in the sky, eclipses, take place when the Sun, Earth and Moon are all in a straight line in space. When the Moon is between the Sun and the Earth, the Sun is eclipsed as the Moon blocks our sightline. Similarly, a lunar eclipse will occur when the Earth is between the Sun and the Moon, and is therefore casting a shadow onto the Moon. Total eclipses of the Moon are commonly seen because the Earth is considerably larger than the Moon, and the Moon is therefore more likely to pass through the Earth's shadow. On the other hand, solar eclipses are seen to be total over only a small fraction of the Earth's surface at any one eclipse. At a given location, a dramatic eclipse of the Sun is a very much rarer phenomenon that an eclipse of the Moon. At any point in history most living persons have never seen a total eclipse.

To make a prediction of a forthcoming eclipse, it is necessary to record carefully the motion of the Moon. The Moon's orbit, or path round the Earth, is tipped at an angle of slightly over 5 degrees to the Earth's annual path round the Sun. This inclination of the orbits results in the Moon's apparent path through the skies being far more complicated than the Sun's. In fact the tilted lunar orbit slews around the Earth, taking 18.61 years to go full circle. An important consequence of this is that the extreme moonrise and moonset positions, as judged against a distant horizon, occur only every 18.61 years.

At Stonehenge the long sides of the rectangle run to the most southerly moonrise and the most northerly moonset. As if that were not enough, one diagonal simultaneously defines two important intermediate rising and setting positions. The satisfying and impressive property about the astronomical alignments at Stonehenge is their beautiful, yet simple, symmetry. One rectangle alone encompasses most of the major rising and setting positions of the Sun and Moon.

All over western Europe you can find ruined stone monuments from the Neolithic Age, and throughout the British Isles there are a great many small circles of stones. These latter are quite adequate, in many cases, for following the seasons, especially when we remember than an error of a week or so is scarcely of any consequence. In fact, a calendar required solely for agriculture can be regulated just as well by reference to migratory birds, hibernating animals and flowering plants. So why is Stonehenge so advanced? Clearly so much trouble was not taken with the lunar sightlines just in order to make a calendar. The intellectual achievement of the architects is the more impressive because so many critical lines are built into the rectangle. If Stonehenge were to be moved to a different latitude, only tens of

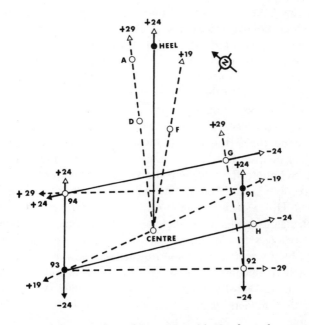

The major sightlines at Stonehenge. Gerald Hawkins demonstrated that the chance of these alignments being accidental is very small. It is more likely that the general plan of Stonehenge arose from a definite desire to follow the motions of the Sun and Moon accurately, possibly in order to predict eclipses.

kilometres north or south of its actual location, it simply wouldn't work! Only at its unique latitude can the Sun and Moon lines be marked by a rectangle. So what was behind this very clever arrangement?

Hawkins and others, prominent among them the cosmologist Sir Fred Hoyle, have speculated that this unique rectangle might have been used to determine the dates of eclipses visible at Stonehenge. An intellectual activity of this order requires a far more accurate device than does a simple perpetual calendar. The eclipse theory also gives a satisfactory explanation for the existence of the 56 regularly spaced holes, which must surely have served some useful function.

One skill that must be mastered in order to forecast eclipses is the ability to keep track of the Moon's leisurely 18.61-year cycle of motion. Three times 18.61 is almost exactly 56, so this suggests that the holes may have served as a counter. If a marker stone is moved along by three holes every year it goes full circuit, almost like the Moon, in 18.67 years. This was one of Hawkins's major findings: he showed how by keeping a tally it would be possible to predict some, but by no means all, of the solar eclipses visible at Stonehenge.

Fred Hoyle worked at the problem and came up with an ingenious, though rather involved, solution. He showed that if four marker stones could be moved, according to a regular scheme and by varying amounts round the series of holes, it would be possible to predict almost every eclipse. His prescription allowed the small errors that inevitably creep into the scheme to be kept under control by making small corrections as the Sun and Moon reached the critical rise and set markers. Hoyle's achievement was to show how we, today, could use Stonehenge as a reliable predictor of eclipses without making any significant changes to the ancient plan. Of course we cannot *prove* that this is what was actually done. But neither can a historian working with written documents *prove* (in the sense intended by scientists) any but the most simple incidents of the past. What can be proved is that modern man could still predict eclipses at Stonehenge today.

Imagine the power the ancient eclipse predictors would have had. The Sun, the light of the world, most powerful of the primitive gods, disappears in a total eclipse. Clearly anyone able to foretell such a drastic event must be able to wield considerable political/religious power. And on the odd occasions when they predicted an eclipse that was not, in fact, visible at Stonehenge, the priestly class could claim that its absence was largely due to their timely intervention in divine affairs. Power indeed!

Almost a thousand years after the ditch-diggers the Early Bronze Age people controlled Stonehenge. These successors had the secrets of metal-smelting, they mined, farmed and traded. Rich burials attest to the stability, wealth and entrepreneurial activity of their rulers, who were able to trade with Scotland, Egypt and Scandinavia. Finally they set to work on a permanent, cathedral-like monument to their celestial knowledge.

The jumble of masonry we see today was assembled only at great cost. For example, the series of stones known as the bluestones, which were finally arranged as a circle and a horseshoe, all had to be transported by boat or raft from the Prescelly Mountains in Wales.

Over eighty blocks weighing 5 tons each were manhandled on sledges and rollers, down to the natural harbour at Milford Haven, thence by sea to the mouth of the river Avon, and overland to Stonehenge. What intense force drove them all the way to Wales to seek these special blocks? We do not know.

Nearer at hand, on the Marlborough Downs in fact, the Bronze Age builders found the massive grey stones that form the arches and circle of giant uprights, or sarsens as they are called. Each sarsen weighs around 25–50 tons and is up to 5 metres (16 feet) long. The builders constructed an outer circle of thirty upright stones, capped by lintels to form a complete ring of connected archways. Within the central area of the monument they also positioned five separate arches, again in a horseshoe pattern. The great central trilithon, or

Total eclipses of the Sun must have been the most feared of all cosmic phenomena to early man. This superb photograph taken by E. E. Barnard on 28 May 1900 shows very clearly the lines of magnetic force emerging from the poles of the Sun. (Royal Astronomical Society, London)

arch, perfectly frames the midsummer sunrise over the Heel Stone.

Some of the skills needed to build Stonehenge have been re-enacted for television audiences. To move a sarsen stone probably required several hundred men, hauling, sliding, positioning wooden rollers, and cleaning the roadway. All holes and ditches had to be excavated with antlers and ox bones. To place the lintels in position would have required large ramps of rubble and earth. Many of the stones have been dressed or tooled, by bashing with boulders and crushing stone against stone. Around 60,000 man-days of effort were put into this aspect of the work alone. The total workload for the whole project must have been around one and a half million working days. Such dedication can only be achieved by stable communities strongly committed to a common cause.

At the finish Stonehenge represented nothing less than the finest scientific instrument in the world; an impressive observatory for the Sun and the Moon. In economic terms, it took more resources (as a fraction of the total available) than any solar observatory of modern times. It is, indeed, in terms of commitment, comparable to the space flight programmes of the 1960s.

The standing stones mark out a profusion of narrow view lines to the horizon, a great many of which coincide with sunrise and sunset positions at critical times of the year. But the huge stone arches do not fix such points to a high degree of accuracy because the central circles of stones are too compact. To mark out a line accurately we need two posts placed a couple of hundred feet apart. So why was the cathedral to the Sun erected at such cost? We don't know but at least the effort has shown posterity that ancient Man had grasped the secret of solar eclipses, and achieved far more in doing so than did the overrated Babylonians. Possibly Stonehenge acted as a centre, rather like the national observatories of our own time, for co-ordinating the work at the smaller circles of stone.

Other indications of early 'solar studies' in Europe come from the burial traditions. At New Grange, County Meath, in Ireland, there is a beautiful passage grave. This chambered megalithic tomb dates from 2500 BC, and is surrounded by a stone circle. The tomb is cleverly constructed in such a manner that on midwinter's day the first rays of the rising Sun strike the burial chamber at the end of the tomb. This critical moment is the turning point of the sun cycle, when the failing Sun, the rays of which warm the burial chamber on this day alone, reverses its downward trend and starts to climb higher in the sky once more.

Gods in the sky dominated the theology of ancient Egypt. The sky goddess Nut supported the vault of the heavens, the Sun being borne across on its daily journey in a chariot. According to the Egyptian cosmology of around 2000 BC, the Sun spent the hours of night journeying through the underworld from west to east. Among the Egyptian temples the complex at Karnak, within the modern city of Luxor, is the most impressive. The fifteen large temples include a group jointly dedicated to the sun-god Ra and the god of Thebes (the ancient city sited where Luxor now stands). The main axis of the Amon-Ra temple marks out the rising point of the Sun on the shortest day of the year to within 0.05 degrees. The architects aligned other temples to the Moon.

An important feature of the Egyptian monuments is the survival of written evidence, in the forms of hieroglyphic or picture writing, and of wall decorations. These leave no doubt as to the intentions of the

The humble sundial is the successor to the giant stone monuments of the Megalithic Age. Sundials show only local time, which often differs from Standard Time. (S. Mitton)

builders and the astronomical achievements of the times. At Amon-Ra archaeologists have found inscriptions about the sunrise as well as references to the horizon. Even the original survey line can be identified, for it has a mural showing a pharaoh assisting Amon-Ra and other gods to lay the foundations. At the top of this wall is a room which must have been dedicated to sun-worship, for a mural includes the inscription: 'Make acclamation to your beautiful face, master of the gods, O Amon-Ra, primordial god of the two lands . . .' This room was what we would call an observatory, a place for viewing the Sun, specifically at that all-important midwinter sunrise.

Six million tons of limestone blocks, aligned to the sunrise on the first day of spring. The sides of the pyramids of Gizeh, dating from 2800 BC, run east-west to within the limits of eyeball measurement. Guarding the pyramids, the Sphinx stares across the desert, gazing at the rosy-pink dawn, able to glimpse the first rays of the Sun at the vernal equinox.

Egypt abounds in the symbols of sun-worship and there are plenty of artefacts for use in systematic sun-watching. Amon-Ra, the pyramids of Gizeh, the Sphinx, and the great temple of Abu Simbel, rescued by UNESCO from the floods of man-made Lake Nasser, stand as the world's greatest sun-sculptures. Twice in the year the brilliant dawn light flashes across the desert to penetrate an inner sanctuary at Abu Simbel where it lights up a statue of Pharaoh Rameses II flanked by two sun-gods.

Much of the art of Egypt lies scattered in museums across the world. In Central Park, New York, and on the Embankment, London, there are twin obelisks. These Cleopatra's Needles weigh 200 tons. Carved from flawless granite, they stand 21 metres (68 feet) high. Originally erected at Heliopolis (City of the Sun) by the ruler who built the Karnak solar observatory, they were capped with gold. They were probably used as vertical shadow 'sticks', for their sides are inscribed to Ra. We don't know exactly what purpose they served, however, because Caesar moved them in 14 BC. The New York obelisk arrived there in 1878 and the other reached London in 1880. Now, moved to a climate that is a good deal less favoured by the Sun, they are deteriorating at an alarming rate and really should be sent back to the desert.

In terms of the breadth of their astronomical knowledge, the Babylonians surpassed the Egyptians, although they paid relatively less attention to the Sun, being instead much preoccupied with the Moon. Mesopotamia, now the country of Iraq, was the cradle of the great civilization that built walled cities 5,000 years ago. And they had their observatories too, the watchtowers called ziggurats, the Tower of Babel being the best known example. Babylonian astronomy centred on moon-watching. They used a lunar calendar, rather than a solar-type calendar as we do. Detailed astronomical diaries were kept by temple scribes who noted the positions of the heavenly bodies by means of cuneiform writing on soft tablets of clay that were subsequently baked. By great good fortune many of these have survived. The British Museum acquired crates of them in the late nineteenth century; they were purchased from dealers in Baghdad, and many were saved from builders who wanted to use these excellent 'bricks' in modern construction! The astronomical diaries enabled the priests of Babylon to make predictions as to the future behaviour of the planets, Moon and Sun. They knew that eclipses came in cycles; one of these lasts for 135 months, during which there are 23 high-risk periods when an eclipse is likely, but not certain. The excellent metonic cycle (named for the Greek, Meton of Athens, who later introduced a calendar based on this cycle) lasts 19 years, and was also known in Babylon. It now seems, however, that the Stonehengers performed better at this particular astronomical game.

In the Americas we find yet more evidence of the paramount importance of sun-worship or, as we should call it, solar astronomy among all ancient societies. Mesoamerican astronomy, before Columbus and before the Spanish conquests, flourished in a culture enriched by fine buildings, art and gold. The Spanish, when they came, destroyed the native American heritage, largely through ignorance and unbelievable greed, and thus caused one of history's greatest cultural losses. Today's knowledge of the astronomy of the ancient Americas is largely derived from the Maya area, where carved stone panels as well as manuscripts known as codexes have survived. Incidentally, the Maya lived in Yucatan, the Guatemalan highlands, and the west parts of Honduras and El Salvador. Their civilization reached its zenith in the period AD 200–900, while European thought lay stifled by the Dark Ages following the collapse of the Roman Empire. After AD 900 the jungle and forest started to encroach on the great cities, so that when the Spanish arrived, in AD 1540, they were already in serious decline.

Mayan astronomy was dominated by numbers, cycles, and, above all, a burning desire to record the passage of time. Nowhere do we find such an obsession with time as among the Maya, who inscribed records of its flow almost everywhere—on stairways, in corridors, panels and walls. So extensive and elaborate was the Maya calendar that it uniquely fixed dates hundreds of millions of years in the past.

Knowledge of the eclipses was excellent. We know that this was so from the so-called Dresden Codex, one of several beautiful hieroglyphic books that survived Hispanic hysteria. In the Dresden Codex there is an eclipse warning table. This table gives an infallible prediction of the 1,034 consecutive solar eclipses that occurred, and were visible somewhere in the world, between AD 206 and AD 647. With only slight modifications we could use this same tabulation to find all eclipses right through to the twenty-fifth century, and probably beyond! The Maya saw time as a co-ordinate that flowed from infinity in the past and onward forever, so their prediction tables carried far into the future.

The same Dresden Codex has the Sun deity, an old god with a large eye. The Aztecs of central Mexico deified the Sun as Tonatiuh, a young red-faced god. He was said to have been created in the ancient capital city of the Aztecs, Teotihuacan, in the highlands. To keep the god out of mischief he had to be fed with the hearts of brave men, which probably had serious consequences for at least some of the natives. This terrible violence and inhumanity caused by sun-worship only ended when Cortes wiped out the whole race of Aztecs.

But the Maya, like the megalith builders of Europe before them, had to keep track of the sun-god. Their dual calendar system used two separate cycles of 260 and 365 days. A year of 365 days had to be watched carefully, because the Earth actually takes 365¼ days to circle the Sun. So the seasons slip out of register by 25 days in a century and slide right through a full year in 1,508 years. To deal with this problem an elitist group of astro-priests followed celestial events, so that the calendar could be aligned to the seasons. They did not use leap days to keep the calendar in register with the seasons, but rather tracked the slippage of the seasons through the 365-day calendar. At Uaxactun, in Guatemala, they constructed a triple temple on a low platform due east of a great Mayan pyramid. At the three critical rising points—the equinoxes, midsummer and midwinter—the disc of the Sun rose above the appropriate roof of the triple temple. Chichen Itza in Mexico, the last great city of the Maya, also has its observatory, with a dome 13 metres (41 feet) high and an internal spiral staircase—a sort of Stone Age Mount Palomar. Through three tunnels in the otherwise solid masonry of the dome the Maya tracked the equinoctial sunset and the setting Moon.

High in the Peruvian Andes the Incas flourished. They regarded themselves as the children of the Sun-god, just like the pharaohs of Egypt. Consequently worship of the Sun figured prominently in ritual and daily life. At its height the Inca empire stretched 1,000 miles from Quito (in Ecuador) to Chile. Machu Picchu, the lost city of the Incas, is set in verdant hills, 2,400 metres (8,000 feet) above the sea, at the edge of the Amazon rain forest. This temple area has a stone gnomon, or shadow stick, that could have marked the passage of time and seasons. The centre of the Universe, in the Aztec cosmology, lay at Lake Titicaca, legendary birthplace of the Sun.

Thor Heyerdahl, in a voyage by balsa-wood log raft from Chile to the South Sea Islands, sought to demonstrate a link between the cultures of Polynesia and the Andes. Could Kon-Tiki, the pre-Inca sun-god, be linked to Tiki, the solar deity of Polynesia? Although he demonstrated that the expedition could be accomplished, this alone does not prove any direct transmission of ideas between the Andes and the South Sea archipelagos.

Further across the Pacific lies Japan, the land of the Rising Sun. In the mid-twentieth century Shintoism, a religion based on sun-worship, asserted itself. Derived from a religious tradition stretching back thousands of years, this faith alleged that the sun-goddess was the founder of the state of Japan, and that the emperor was descended from the Sun, and furthermore that the emperor's family would rule Japan for ever.

Clearly, then, since earliest times sun-worship has been widespread among a variety of peoples and cultures. In our own time this obsession takes a less religious form: the beaches of the Mediterranean, California, Florida, and Sydney swarm with oiled worshippers. Not so long ago, when most people lived in peasant communities, the smart set deliberately avoided the Sun, a tanned skin being the sign of a rural labourer. Now roles are reversed and commercial man, his skin a ghostly pallor from office or production-line incarceration, longs for the burning rays and the hot sands.

Perhaps, but not certainly, as long as 20,000 years ago, systematic observations were made of the Moon's phases. This is suggested by the scratches on carved bones. But without doubt, knowledge of the Moon and Sun had matured independently in the Americas, in the Middle East and in western Europe by 3000 BC. The uninterrupted experiences continued until the time of the Spanish conquest in the Americas, but in the Old World were submerged rather earlier by new migrations, new philosophies and new religions. How fortunate then for us that the megalith builders in particular constructed enduring monuments from which we can construe so much about their beliefs. It is hard for us to grasp the huge significance the sky had for the inhabitants of the ancient world. But next time you see a beautiful sunrise in a quiet natural setting look around you and observe nature. I remember seeing glorious dawns in Australia. From Siding Spring Mountain I could see thousands of square miles across the grazing lands of New South Wales. The stars vanish one by one in the lightening sky. From a high point the eastern sky is light grey, then tinged with pink, rosy. Finally a brilliant blood-red Sun rises, seemingly out of the ground, steadily, inexorably, frighteningly large. All around the air explodes in a cacophony of sound, shattering the morning stillness. Splendid parrots squawk, and the shadow of the night moves swiftly over the plains below as dawn marches westward to the great central deserts of that continent. Nature taught our ancestors to worship the Sun. They were closer to Earth and sky than we are, but in our own hearts we still praise the dawn Sun.

Our Nearest Star 2

There are many centres of modern sun-worship; Miami, Bondi Beach, Costa del Sol, to name but three. But how many of the sun-seekers actually realize why the Sun generates the ultraviolet rays responsible for darkening the skin? It takes about a million years for energy released in the heart of the Sun to reach the golden surface, and only a further eight minutes to reach the Earth. The warm rays that we enjoy today were created long before Homo Sapiens had emerged as a clearly identifiable species. Who will be here to enjoy today's energy release when it finally breaks free of the solar interior one million years hence?

Our Sun is an example of a star. Because it is our nearest star it has a particular value for astronomers. Planet Earth is, on average, 150 million kilometres (93 million miles) from the Sun; the next nearest star, called Proxima Centauri, is a feeble spark of light 4 light years, or 40 million million kilometres from Earth. So the Sun is a quarter of a million times nearer to us than the nearest star. We can put this range of distance into everyday language. You're probably holding this book about 30 cm (12 in) away from your eyes. Suppose the next nearest book was as far away as 75 km (50 miles), what could you see? Even a powerful telescope would reveal little more than the size and colour of the book, and nothing of its content. If you could read the title on the jacket you might classify it as a novel, travel book, dictionary, or whatever. An astronomer is in much the same position. The Sun can be studied in detail, but when it comes to the stars we may have to satisfy ourselves with measures of only the colour, size, type and approximate composition.

Our Sun then is the most important star in the sky because it's the nearest. Of course it is ultimately the source of all the energy used on Earth, with the exception of the energy manufactured in nuclear power stations or released in radioactive decay. So we all have an interest in the Sun because it sustains the light and heat that plants and animals must have in order to live. Over a century ago (1871) Richard Proctor chose as title for a book on solar physics: *The Sun: Ruler, Fire, Light and Life of the Planetary System.*

But the astronomers' interest goes far beyond everyday life: here is a cosmic laboratory in which we can understand processes that are of importance throughout the Universe as a whole. The Sun can teach us about the Universe itself, as we shall see, provided we can untangle the mysteries of its birth, life and ultimate death. Today about one-fifth of the world's professional astronomers, maybe 500 people in all, are principally engaged in solar research. A large number of amateurs specialize in solar observation.

Stars are not the same as planets. A star, like the Sun, is a glowing globe of hot gas with a fiery nuclear furnace. Stars, being entirely gaseous, have no permanent surface features. Planets are lumps of rock or balls of cold gas. The major distinguishing feature is that energy flows out of stars—as heat, as light, and as other radiations— energy that was once trapped inside the central nuclei of atoms. Planets, in contrast, have negligible sources of internal energy, and rely on nearby stars for their main supplies. A star shines of its own accord, but a planet will only show in our optical telescopes because it reflects sunlight. One of the many fundamental facts that intrigues me about the Universe is that the laws of physics require nuclear reactors the size of the Sun in order to extract energy from the sub-microscopic world of the proton and neutron. Twenty-two powers of ten separate the size of the atomic nucleus from the dimensions of the solar reactor needed to release its immense energy.

I have already mentioned that the Sun is about 150 million kilometres from the Earth. To find the distance to the Sun you have to be able to think your way through geometrical ideas. So among the astronomical communities of the past the Greeks alone managed a sensible crack at the problem, for they were supremely good at what we still call Euclidean geometry. Over the entrance to the Academy in Athens was written: 'Let no one enter here who knows no geometry'. Aristotle (384–322 BC) studied the work of Plato (c.427–347 BC) at the Academy (founded 287 BC), and subsequently had a profound influence on solar studies for almost 2,000 years. One of Plato's followers, Euclid, who lived in the fifth century BC, gave the first coherent description of the geometry of straight lines and perfect circles. He showed how to derive the main results, logically and systematically, from a single set of assumptions.

With this useful geometrical heritage, Aristarchus of Samos, one of the giants of Greek astronomy, attempted to measure distances. Before this time there were only vague guesses: Anaxagoras (500–427 BC) taught that the Sun was a blazing stone about 50 kilometres (30 miles) across. The principle of Aristarchus' method is absolutely correct, but difficult to apply. He argued that when we can see a precise 'half-moon' in our sky, the angle between the directions from Sun to Moon on the one hand, and Earth to Moon on the other, must be exactly a right angle. In this situation a measurement of the angle between the directions from Earth to Moon and Earth to Sun then gives all the angles in the Sun-Moon-Earth triangle.

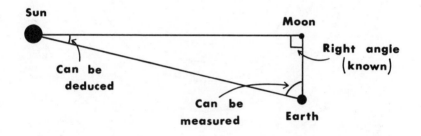

Aristarchus understood the geometry of the Sun-Moon-Earth triangle. He realized that when we see a half Moon the triangle has a right angle in it, and measurement of the Sun-Earth-Moon angle will allow the relative distances to the Sun and Moon to be deduced.

Aristarchus, who lived from 310 to 230 BC, measured the vital angle Sun-Earth-Moon as 87 degrees. From this he concluded that the Sun is about twenty times further off than the Moon, and quantified this at about 6 million kilometres (4 million miles). Now before we pooh-pooh this as a big failure we have to keep in mind the following: Aristarchus was the first astronomer of consequence to reason that the Earth went round the Sun and that the Moon shone by reflected sunlight; and he did at least show that the scale of the solar system is enormous by terrestrial standards. The Sun indicated that the Universe

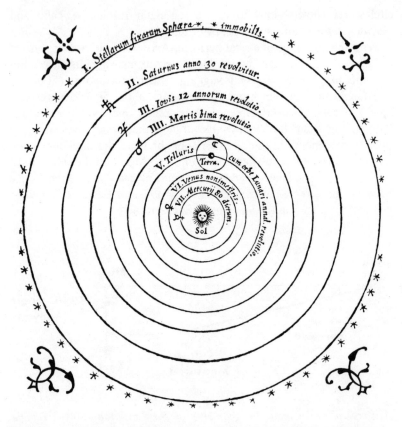

This famous diagram published by N. Copernicus in 1543 established the Sun as the centre of our solar system.

is much larger than the Earth, a significant result in its day. The difficulty in this observation, one of the very few genuine observations made by the Greek philosophers, lies in guestimating when the Moon really is at the halfway stage. The distance depends crucially on getting the angles right, and Aristarchus' 87° lay sufficiently far from the true value of about 89°85 to throw his calculations right out.

Two millennia elapsed before the new age of the scientific method, dating from the sixteenth century, began to contribute to our knowledge of the scale of the solar system. The story begins with Copernicus (1473–1543), who, in a book withheld from publication until the year of his death, obscurely set out reasons for believing the Sun to be at the centre of the solar system, rather than orbiting the Earth. (Aristarchus had been ditched by this time in favour of Aristotle, whose influence was paramount from the thirteenth century onwards. Aristotelian philosophy was Earth centred.) At first Copernicus' ideas found favour with the Catholic Church, not being held erroneous until 1616. But these ideas then became part of the controversial intellectual turmoil that was to sweep through seventeenth-century Europe.

Tycho Brahe (1546–1601), the first modern observer, set about accumulating data on planetary movements at a lavishly equipped observatory off the coast of Sweden. Johannes Kepler (1571–1630) worked briefly as Tycho's assistant and inherited all the data upon the death of his master. Kepler, the mystic and astrologer, devoted fantastic labour to the analysis and interpretation of Tycho's observations. Finally, after nearly two decades of slog, he came up with the answers: he found that planets orbit the Sun on elliptical paths; discovered the relationship between the size of the orbit and the time for the planet to orbit the Sun; and discovered how the velocity of a planet varies along the orbital path. This work established the Sun as the unchallengeable ruler of the solar system. Isaac Newton (1642–1727), a brilliant English mathematician, was to discover the root of the Sun's regal authority.

Kepler had stumbled across a set of rules. What Newton showed was *why* planets circuiting the Sun do so in the observed manner. After developing the branch of mathematics known as calculus, Newton was poised to unify the forces of Earth and sky in one brilliant synthesis: the theory of universal gravitation. The story of how the fall of an apple led Newton to this theory is well known. He showed that the force that pulls a falling object to Earth also holds the Moon in its orbit around the Earth. By extending this idea he could show that the gravitational force of the Sun held the planets on their orbits. Furthermore, the shapes of the orbits—ellipses—arose naturally in the Newtonian theory, whereas all Kepler had in essence said was: That's what it looks like.

The dynamical laws formulated by Newton opened anew the possibility of *calculating* the Sun-Earth distance. Even today, in the late twentieth century, direct measurement of this distance is difficult. Instead the technique is to relate it to distances that can be determined with some ease. The basic approach is to find the distance at a particular time from the Earth to another object orbiting the Sun. If this can be done for one distance in terms of terrestrial units then Newton's laws permit us to compute all other distances because ratios between them are known to great accuracy.

Astronomers first attempted to measure the distance to Mars, since it comes closer than any other planet. The method is to measure its apparent place in the sky from a variety of locations on the Earth. Because of parallax the nearby planet does not appear in precisely the same location when viewed from different observatories. If the distance between a pair of observatories is known, and the difference in the angular position of Mars is measured, then its distance from Earth at the instant of the observation may be calculated. Telescopes positioned on opposite sides of the Earth will differ in the angle at which they have to point in order to observe Mars by almost $0''\cdot75$. The first application of this method gave a distance of about 85 million miles.

Captain James Cook, on his first voyage of discovery (1768–71), rediscovered New Zealand and charted the eastern coastline of Australia, making landfall at Botany Bay in 1769. As an explorer, Cook was primarily motivated by a quest for knowledge: geography, peoples, and science. In 1769 he carried out a most important observation. At rare intervals, only twice per century, the planet Venus is seen to pass across the face of the Sun, as viewed from Earth. This is known as a transit of Venus; it is rare because the Earth and Venus lie in a pair of orbits that are tilted with respect to each other, so the Earth, Venus and the Sun hardly ever lie along one straight line. Cook observed and timed the transit from a far-southerly location. These data were helpful, when combined with measurements from the northern observatories, for refining knowledge of the dimensions of the solar system.

In 1877 the Scottish astronomer Sir David Gill (1843–1914) set off with his wife to lonely Ascension Island in the South Atlantic. The island had no town, just a naval garrison. It was here that Gill made a series of classic observations of the planet Mars which were needed for finding the scale of the solar system and the distance to the Sun. He observed Mars when it was at its closest approach to the Earth for a century. The method involved observing the position of the planet relative to background stars in the early morning and evening. The principle used to reduce the observations was that the Earth's rotation had displaced the observatory by a known amount between observations. Laborious calculation then yielded an accurate distance of 93 million miles, at a time when most estimates varied from 90 to 95 million miles. Gill chose Ascension Island, incidentally, on account of its good weather and southerly position. In 1931 a very close approach to Earth (25 million km, 16 million miles) of the asteroid Eros enabled astronomers to gauge the scale of space to within 0.01 per cent.

Until comparatively recently the use of sightlines to the planets provided the only means of finding the Sun-Earth distance. The mean value of this distance is so important in modern astronomy that it has its own name: the astronomical unit. Today, however, the method in use has phenomenal accuracy. Planetary radar is used to pin down the value of the astronomical unit to an uncertainty of a few kilometres!

The radar method is easy to understand. Pulses of radio waves are beamed from a transmitter on Earth towards the planet Venus. The radio waves are partially reflected by the planet's solid surface. Although the returning signal is very weak after the round trip of 100 million kilometres, a sensitive radio telescope can detect it. In practice the same telescope is used both as radar transmitter and radio receiver. When the elapsed time between the transmission of a pulse and its receipt back at the telescope is measured, the distance to the planet is immediately known: radio waves travel with the velocity of light, which is known with a precision of one part in a million million. By means of the radar technique, distances within the Sun's retinue of planets are pinned down to an accuracy of one part in a few hundred million. That's equivalent to finding the distance between a point in London and a corresponding point in New York to within a few centimetres. But now that spacecraft are sent to the outer reaches of the solar system, such accuracy is vital to the success of long-haul missions which may take ten years or longer.

The apparent diameter of the Sun is about 32 minutes of arc, slightly over half a degree. In fact, it varies a little as perceived by astronomers on Earth. This is because the Earth, moving as it does on an elliptical orbit, is not always at the same distance from the Sun. When we are nearest, and curiously this is in January, the coldest month of the northern winter, the Sun's disc is 32½ minutes across, whereas in July, when we are at the greatest distance away, it is 31½ minutes.

By combining information on size with distance we can work out the solar diameter. The Sun's diameter is 1.4 million km (865,000 miles) in round numbers. This is 109 times the size of the Earth. In terms of volume then the Sun is 1.3 million times greater than the Earth.

Newton's law of gravitation allows us to weigh the Sun. Earth is trapped in a gravitational path that takes it on a perpetual orbit about the central force of the Sun. The speed of our planet's journey and its distance from the Sun are just two of the quantities that determine the orbital path. The Sun's mass also enters into the calculation, so that measurements of the quantities associated with the Earth's motion (speed and distance) allow the solar mass to be found. In fact the mass is one-third of a million times that of the Earth—2,000 million million million million tons (1.989×10^{30} kilograms, to be precise).

As an illustration of how the method works in principle we can ask what would be the effect of a different solar mass? If the solar mass were doubled, the Earth would have to move twice as fast to maintain its present orbit; similarly, if the mass were half its actual value, the velocity of the Earth would have to drop by the same factor to stay in orbit.

The relation between the Earth's mass and the Sun's mass is such that a scale model with an Earth weighing 10 grams (0.33 oz) would need a 3-ton sun for realism. But the difference in mass is smaller than that for volume. This shows that the Sun is made of material that is, on the average, less dense than the rocks of our planet. The Earth has an average density of 5.5; in other words a bucket of typical Earth-stuff is over five times as heavy as a bucket of water. But the Sun's relative density is only 1.4. Now consider: the tiny solid Earth packs four times as much matter per average volume as the giant fiery Sun. Why? We have a hint here that the Sun is made of gas, not solid matter.

At the surface of the Sun the crush of gravity is rather higher than we are used to on the Earth. In fact, gravity pulls twenty-eight times harder, a result of the much greater solar mass. So a typical human would weigh over 2 tons close to the Sun. Furthermore, this fictitious explorer would need a powerful rocket to get away because the velocity needed to escape the Sun's gravitational embrace is over 600 km per *second*—1.4 million miles an hour. Rubbish thrown out of the rocket would crash down to the solar incinerator at an impressive rate: in solar gravity objects fall almost 150 metres in the first second of free fall, compared to 5 metres at the surface of the Earth.

In the last hundred years especially, the Sun has frequently featured in investigations of considerable interest to physicists. A very important effect of the Sun's gravity contributed to the vindication of Einstein's general theory of gravitation in the early part of the twentieth century. This famous theory substituted curved space for the

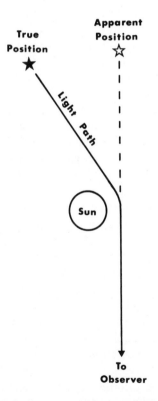

The deflection of starlight as it passes close to the Sun was found, in 1919, to accord exactly with the prediction of Einstein's General Theory of Relativity. The success of this observation, masterminded by Sir Arthur Eddington, firmly established Einstein's scientific reputation.

straight-line geometry of Euclid and Newton. Einstein's work showed that the shortest distance between two points is *not* a straight line when the effect of gravity is correctly included in the theory. He asserted that a ray of light, for example, would follow a slightly curved path if it passed close to a substantial mass. Now the differences between the earlier theory of Newton and that of Einstein are imperceptibly small at the level of everyday life on planet Earth. When you drive your car safely along the highway you are subconsciously using the Newtonian laws of dynamics, because the modifications called for by the theory of relativity are ridiculously small. Only over vast distances, or long timescales, or in situations where very massive objects or high velocities are encountered, is it necessary to abandon Newton's theory. Consequently the test of the General Theory of Relativity had to proceed via observations of the effect of the Sun because its gravity is high enough to give measurable phenomena.

Einstein calculated that the gravitational bending of a ray of light passing close to the Sun would be 1.7 seconds of arc deviation from a straight line. The first opportunity to test this prediction arose in 1919, at a total eclipse of the Sun. Photographs taken while the eclipse was in progress showed, on later analysis, that the light from a star whose sightline almost grazed the Sun had indeed been pulled off course by the predicted amount. The expedition was organized by Eddington immediately after the 1914–18 war. The result immediately brought international acclaim to Einstein.

A further solar puzzle was resolved by Einstein's theory. The long axis of Mercury's elliptical path swings slowly but surely in space. This is an intriguing situation: not only is the planet rotating around the Sun, but the orbit is as well, by 43 seconds of arc per century. In the general theory of relativity this rotation is a natural consequence of Mercury's changing velocity as it pursues its eccentric orbit.

So, we see that solar studies assisted theoretical physicists seeking a confirmation of what at the time was the most difficult theory ever proposed. And this theory has stood the test of time, being generally accepted as a correct description of the relation between matter, gravity and the structure of space and time.

It is true that in our own time we have to go beyond the solar system for real tests of the theory, but this does not diminish the importance historically of the classic observation of 1919.

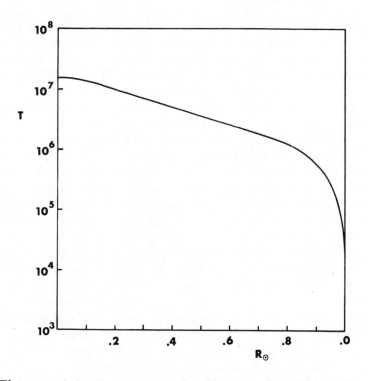

The vast variations in temperature through the Sun provide many different physical regimes.

Sun-watching will certainly continue to make major contributions to physics, as well as to astronomy, because the Sun is an arena in which phenomena that could never take place in a terrestrial laboratory can be seen to happen. The examples drawn from the theory of relativity give one illustration, and we shall meet several more.

Another example of the astrophysical laboratory is provided by the variation of temperature through the Sun and its atmosphere. Measurements of the energy radiated by the Sun show that the temperature at the surface is a searing 6,000 degrees Centigrade. This exceeds the temperature at which all known materials will melt. To put it another way: we cannot yet build a space capsule that would not be turned into a puff of gas long before it reached the golden solar surface. The transparent outer layers of the Sun curiously have a higher temperature still, running into millions of degrees. And, as we penetrate the solar interior, we encounter an ever rising temperature and pressure, until, in the central region, the conditions result in a temperature of millions of degrees and the pressure is thousands of millions of times higher than on the surface of the Earth. You can already see that the inside of the Sun can only really be explored by mathematicians and computers since no instruments could conceivably penetrate directly even as far as the solar surface.

What then are the main facts that we want to know about the Sun? Since it profoundly affects everything that grows and moves on the Earth, as the ultimate energy source, it's important to know as much as possible about the central power station that feeds the Sun. Also we want to find out if the energy transmitted by the Sun is constant or not. Any changes in the Sun's energy emission would seriously affect the climate—and weather—on the Earth. Could this be the cause of climatic variations, particularly the great ice ages? As a highly evolved animal, Man is particularly dependent on the Sun staying much the same. The energy sources that our technology has released have, to a great extent, obscured the fact that the Sun is the only completely reliable and environmentally safe energy source. Fossil fuels will not last forever and nuclear energy may be too hazardous on a really vast scale. Presumably cheap energy supplies will eventually be exhausted. Maybe this will be in the far future (let's hope so!) but the problem will have to be faced by someone, somewhere, at some as yet indeterminate time in the future.

When people venture into space, the behaviour of our Sun is of major interest. This is because the Earth's atmosphere is a marvellous shield against harmful forms of radiation from the Sun. In other words, we have evolved from lower mammals and the forerunners of mammals in the absence of damaging radiation. If instead there had been a strong influx of ultraviolet rays received at the surface of the Earth, for example, then, in a real sense, we would have evolved with very thick skins! Astronauts are not super-persons, so the spacecraft have to be equipped with adequate shielding from high-energy solar rays for the protection of the flight crew. When astronauts need to leave their spacecraft, to explore the Moon or to carry out assembly tasks, ground-based astronomers keep a careful watch on the Sun. They are looking for sudden energetic outbursts that would spew deadly radiations into space—radiations that are stopped by the spacecraft shielding, or our atmosphere, but not by a regular spacesuit. Further understanding of the Sun and its radiations is an essential part of successful manned space-flight, as well as a major contributor to weather forecasting and climate modelling.

Astronomers, like other scientists, want to find out as much as possible; but the acquisition of knowledge for its own sake is selfish, particularly if it involves the expenditure of public money. Just sitting at a telescope and collecting data is a pointless occupation unless the observer's aim, perhaps only in the back of the mind, is to gather data that will assist in the solution of real problems. Here we come to a genuine dilemma for the modern researcher: to make a given research effort feasible, to convince a responsible committee to allocate scarce public funds, one must seek to solve a problem that is small and therefore apparently capable of a quick solution. Hence modern research, and solar physics is no exception, usually consists of a vast series of apparently insignificant investigations. But the sum total is, we hope, increasing our ability to reveal important truths.

The Northern Lights over Alaska. These beautiful displays are the result of collisions between charged particles from the Sun and atoms in our atmosphere. (S-I. Akasofu, University of Alaska)

There are still many problems to solve. For example we want to comprehend, in detail, the basic operation of the Sun's energy source. This is crucial for an understanding not only of the Sun but the stars as well. Theoretical work can show, for example, how long the Sun can continue to function in the present way.

What is the Universe made of? Much of the visible matter is in the form of stars, and most of those are like the Sun. The Sun's composition can provide decisive information on the history of matter in the Universe as a whole.

Although Aristotle rashly asserted that the Sun is without any kind of blemish, we know that this is quite untrue. Dramatic outbursts of energy are frequently recorded by telescopes. We want to know what triggers these bursts of energy, and how they influence the Earth, climate and life.

The link between Earth and Sun goes beyond the behaviour of the climate. All the time, a wind of atomic particles is streaming out of the Sun and brushing past the Earth. This breeze of solar electricity is responsible, among other things, for the auroral displays, the moving veils of light that are so prominent in the polar regions. Measurement of the wind from the Sun is important for radio communications. This is because the electrically-charged particles in the solar wind interfere with layers of our own atmosphere that are responsible for the reflection of long-distance radio waves. If the Sun alters the structure of a layer, radio communication over long distances is liable to fade out.

Finally, studies of the Sun can throw light on some of the questions that intelligent people have always asked themselves: where did the solar system come from? How old is it, and what materials are found in it? Theory and observation combine here to give estimates of the age of the Sun. Sunlight also carries important information on the composition of the Sun's outer layers. These regions still have the basic chemical elements in much the same proportion as the material from which the Sun and infant planets formed. The planets have become drastically modified since, because those nearer to the Sun have lost much of the light-weight gas they once possessed. How and from what did the Earth form? To answer this question we have to move our attention away from home to the centre of the solar system.

Probes for Our Nearest Star

3

The first observers came with wood and stone. They set up sight-lines and observatories to keep track of the annual cycle of sunrise and sunset. All that they could measure were angles, and yet from these crude data they probably deduced useful information about the seasons and eclipses.

Contrast this simplicity with the complexity of a modern solar observatory. Telescopes can see the Sun in a variety of ways: visible light, infrared heat rays, invisible ultraviolet and X-rays, and the radio waves. Without telescopes we should know very little about the Sun other than its position in the sky. On mountain tops, where they can take advantage of clearer air, optical telescopes watch the Sun continuously. As the Sun sets on one continent it rises over another, so that the world-wide sun-watch is uninterrupted by the natural cycle of day and night. Similarly, the heat radiated by the Sun is monitored, particularly at meteorological stations. In orbit about the Earth there are satellite-borne instruments to monitor unexpected bursts of high-energy radiation. And in Australia an array of ninety-six radio tele-scopes is able to make radio images of the Sun, and in doing so pinpoint regions of unusual activity.

The aim of all facets of sun-watching, of which the above are just a few examples, is to increase the fund of reliable information about our daytime star. This data bank can then act both as a stimulus and as a check for theory. Surprises in the data lead to new models of how a particular aspect of the Sun works but the existing information limits the variables that scientists can put into the models. This is how science in general should proceed, with observation and theory assisting each other. The purpose of theory, of course, is to explain what is already measured, in terms of physical laws, and to give sound predictions for the future. Several problems of solar astronomy are *only* accessible by theory. For example, how long will the Sun last? Yet the answer to this question is only reliable if supported by

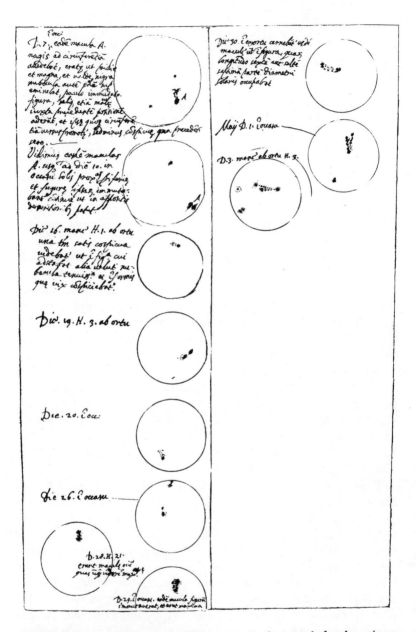

Galileo's observations of the Sun provided the first good sketches of sunspots. *(Yerkes Observatory, Wisconsin)*

good data. The quest for the best quality information, and lots of it, has led to the construction of telescopes designed specifically for observing the Sun in all its manifold aspects.

Galileo was the first person, so far as we know, to turn a telescope sunwards. This is actually an extremely foolhardy experiment that you must not attempt to repeat under any circumstances. But you can make your own observations of the Sun using a telescope in the following completely safe way.

The aim is to set up the telescope so that, instead of forming a focussed image at the normal viewing position or distance, it makes one further away from the eyepiece. This can be achieved by pulling or racking the eyepiece out beyond the normal setting for astronomical viewing. Incidentally, it's easier to get set up for Sun observations if you use the eyepiece of lowest magnifying power (assuming you have any choice). So much for the eyepiece, now what about the telescope? For complete safety I like to remove the small finder telescope that sits piggy-back on the main tube. The reason is to avoid the temptation that any uninformed spectators might feel to look through the finder even for a moment. With the finder off, it is now necessary to make a shield to fit round the front end of the telescope. The aim here is to cast a shadow behind so that the solar image can be discerned by projection onto a screen. This simply does not work unless the screen is shielded from the direct light of the Sun. Your shield does not need to be in the least bit fancy. I sometimes use a piece of cardboard (an old detergent box will do!) and cut a hole of a size that fits snugly over the tube. It is essential, of course, that the telescope is on a stand or mount of some sort because the whole purpose of this exercise is to remove the need to look along the barrel of the telescope towards the Sun. A hand-held telescope cannot be aimed at the Sun and kept steady without sighting along the tube.

The telescope with its Sun shield can now be aimed at the Sun by trial and error. A white card held a little beyond the eyepiece is used as projection screen. Through patient experiment it soon becomes easy enough to focus a sharp image of the Sun onto the card. If this is a few centimetres in diameter then one or more sunspots may possibly be seen if the image is well focussed. With a few minutes' practice it is easy to distinguish the sunspots, with their characteristic penumbras, from the images of specks of dust in the telescope.

It is very dangerous to look directly at the Sun. The author demonstrates a safe method of viewing sunspots by projecting an image using a simple telescope. (J. Mitton)

Sun-gazing in this manner is safe and fun. By making observations over a period of time it is possible to see the changing pattern of markings on the face of the Sun, which Aristotle so erroneously stated to be without blemish. This is also a good experiment for school science teachers, for it is one of the few astronomical demonstrations that can be made during normal hours. But never forget that nobody must be permitted to look through the eyepiece directly at the Sun. If you give a demonstration be sure to make your audience understand that they must not try to look through any instrument they might happen to have at home.

Now I have another warning. Inexpensive telescopes sometimes come with small Sun filters that are intended for use in conjunction with the eyepiece. The assumption is that the dark glass cuts down the radiation from the Sun. Although this is partially true, the filters are very dangerous. A telescope lens, even of the smallest type, collects at least a hundred times as much light as the unaided eyes. So even if the filter absorbs 99 per cent, the light is still blinding. Furthermore, if the filter shatters, as it would do with all that energy focussed on it (a piece of broken glass will cause a fire, remember), then your eye is completely unprotected. Finally, the cheaper filters are not effective at cutting out ultraviolet light, the most damaging radiation. British amateur astronomer Patrick Moore, noted for forthright commonsense, recommends dropping these filters far out at sea! Perhaps it's easier just to throw them away, but in any case they should not be kept with the telescope for fear that an unwise observer should dare to use one.

An entirely different type of filter, an interference filter, is available for use with the more expensive amateur telescopes. These filters fit over the front aperture of the telescope and use the physical properties of light to reduce greatly the transmission of radiation. Waves of only one very narrow band are allowed through, perhaps only 0.1 per cent of the sunlight. We shall be meeting interference filters a little further on because they have major professional applications as well. The ones made for amateurs cost several hundred US dollars. A keen observer would not begrudge this for they bring a rich variety of high-energy phenomena into view.

Many telescope owners will want to try solar photography. This is a specialized undertaking, but very rewarding if done successfully. The main problem, of course, is dealing with the tremendous dazzle of heat and light. Special tricks are used to overcome this but it is not appropriate to give all the details here. If you are interested in this aspect it will be necessary to order handbooks on astronomical photography from your library or bookstore. Recommendations are given in the bibliography at the end of this book.

Professional observatories require many types of instrument for solar work. Of course, all of these will not be found at any one observatory because each will generally specialize in some aspect of research and will have instruments designed for the job in hand.

One of the obvious quantities to measure is the amount of energy the Sun sends out. In practice this is accomplished by finding how much energy is received at the Earth and then applying appropriate scaling factors to get the output or luminosity. The amount of energy received per square metre at the top of our atmosphere is called the *solar parameter*. It used to be called the solar constant until astronomers and weathermen suspected that it might, in fact, be varying a little. The definition is referenced to the top of our atmosphere because the transparency of the air is not the same in all parts of the world. If the measurements are made at ground level a correction must be made for the energy which has been soaked up by the atmosphere. Both astronomers and meteorologists need to measure the strength of incoming solar radiation. The principle of making the measurement is to use the incident energy to warm an object whose temperature can be accurately measured, or to release electrons from a semi-conductor which can then be counted. For the technically minded, the instruments have various names: thermopiles, bolometers, radiometers and pyrheliometers; only the basic principles, as opposed to constructional details, need concern us.

As long ago as 1837 C. S. Poulet measured the Sun's intensity in the following imaginative way. He took a copper pot, painted it black so that it would reflect as little sunlight as possible, filled it with water and stuck in a thermometer. First the black pot was placed in the shade, and the temperature read. Then he put it in the sunlight and measured the rise in temperature per minute. He made a correction for the atmospheric absorption and got to within 10 per cent of the right answer. I find this fascinating; using ordinary domestic equipment and no money, Poulet got an answer that is good enough for everyday purposes. Improvements on this homespun method have required us to capture the Sun's radiation across the whole of the spectrum from infrared to ultraviolet. The spectrobolometer, for example, is a device also perfected in the last century for measuring the distribution of energy content across the spectrum.

At the standard distance of one astronomical unit (93 million miles) from the Sun, the energy flow is 1.36 kilowatts per square metre, or 1.16 kW per square yard. Not all of this gets down through the atmosphere. If the Sun is directly overhead, about one kilowatt of power hits the ground for every square metre at right angles to the Sun's rays. By calculating the area—in square metres or yards—of a shell with the radius of one astronomical unit, we can work out the total amount of solar energy streaming out into space from the Sun. It comes to an astonishing 3.83×10^{26} watts, and of this 2×10^{17} watts are intercepted by the Earth. The mind cannot grasp such numbers: 10^{26} is a 1 followed by 26 zeros—100 million million million million. But consider the following: the amount of solar energy actually reaching the Earth's surface is at least 10^{14} kW, whereas the total of all man-made power of every type is only 3×10^{9} kW. The solar-energy input to the Earth is around 30,000 times as high as the artificial input.

Now we can ask, how does this impressive performance for the daytime star compare to the celestial competition? At this point I need to sidetrack for a discussion of *magnitude*, because astronomy is a science that still uses a few ancient concepts. Thus, the energy outputs of stars are not expressed as calories, or watts, but as magnitudes, a hangover from twenty-five centuries back when the Greek philosopher Hipparchos set out a relative scale for comparing star brightnesses. (Astronomers are not the only ones dragging their feet: the performance of internal combustion engines is usually expressed in horsepower rather than kilowatts.) On this ancient magnitude scale the brightest stars are assigned to the first magnitude, those that are only just visible to the naked eye to sixth magnitude and the intermediate-brightness stars to magnitudes between 1 and 6. Of course, the system was modified to take account of modern scientific methods in the nineteenth century, so the apparent brightness can be specified to within as little as one-hundredth of a magnitude interval.

You probably noticed that the magnitude system gives faint stars higher numbers than bright stars. The faintest objects that have been glimpsed by telescopes have magnitudes of about +26. The brightest star in our sky, Sirius, has a magnitude of −1.42, while planet Venus

The heliostat atop the McMath solar telescope. The mirror tracks the Sun in order to deflect an image down the solar tower. In the background are several other astronomical telescopes at Kitt Peak. (Kitt Peak National Observatory, Arizona)

reaches a maximum brilliance of −4.4 magnitudes. Our Sun has an apparent magnitude of −26.7 magnitudes. In all more than 52 magnitudes, or 10^{21} in terms of relative energy received at the Earth, separate the Sun from the faintest galaxies. When we look at the Sun we are seeing as much light as we would get from ten thousand million stars like Sirius. This difference, however, I must stress is due entirely to the proximity of the Sun, and is not the result of any super-power it generates.

Astronomers, of course, are mainly interested in measuring solar energy with a view to seeing how the Sun works. Meteorologists, on the other hand, follow the Sun as the major influence on climate. For this reason the Sun's performance is measured on a daily basis at many thousands of weather stations throughout the world.

In order to monitor the solar surface, or to watch the 'weather' on the Sun, a telescope of some sort must be used. Sun-watching telescopes are usually quite unlike any other species of astronomical telescope. There are two principal reasons for this. First, the Sun tracks through only a fixed range of altitude and horizon angle, so there is no need to be able to reach every part of the sky as with a conventional astronomical telescope; and second, the brilliance of the Sun means that a telescope is needed to produce fine images, and not to gather a great deal of light. The first consideration means that the main telescope structure is often permanently fixed, and a moveable mirror, termed a heliostat, is used to reflect the sunlight into it. The second consideration requires the use of an imaging mirror or lens of long focal length because these produce large images that are not overpoweringly brilliant.

One of the greatest of modern solar telescopes is the McMath solar telescope at the Kitt Peak National Observatory, Arizona. This is used by scientists to study the Sun throughout the daylight hours, and it can even be used for some stellar research at night. Crowning a 35-metre (110-foot) high tower is the heliostat; this tracking mirror is 1.5 metres (5 feet) across, and its job is to follow the Sun automatically in order to catch the solar image and reflect it down through the main path of the telescope. The main axis of the telescope runs parallel to the rotation axis of the Earth. The focal length is 90 metres (300 feet), so three-fifths of the telescope is actually below ground. An advantage of the scale of the telescope is that it makes an image of the Sun which is 90 cm (3 feet) in diameter. This allows the solar astronomers to see much detail on the surface, and to see how it changes as the day progresses. In order to keep the Sun's image steady, the air inside the long shaft is cooled. This is accomplished by pumping cooling water through tubes that run round the circumference of the shaft. This beautiful telescope allows solar astronomers to study magnetic fields, motions, and the atmospheric composition in detail.

Several observatories possess solar towers. Essentially these are similar to the Kitt Peak solar telescope, except that they are constructed with a vertical optical path rather than one at an angle. In the United States there are solar towers at the Mount Wilson Observatory and the Big Bear Observatory. The latter is situated on an island in a small mountain lake. The planners of the observatory chose this site because solar observations have to be made mainly around midday, when the heating effect of the Sun's rays is at its strongest. By placing the telescope over a large body of water, such as a lake, the image shimmerings due to rising currents of warm air are considerably reduced.

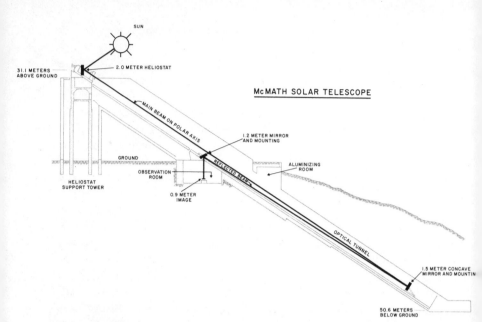

The general arrangement of the McMath telescope, half of which is inside the mountain. The tunnel is cooled in order to help produce a steady image. (Kitt Peak National Observatory, Arizona)

The image of the Sun is formed on a table top inside the observation room of the McMath telescope. (Kitt Peak National Observatory, Arizona)

Apart from straightforward photography of the solar surface, a major activity is solar spectroscopy. This technique aims to read the message of sunlight, for this intense radiation carries within itself important information on the temperature and composition of the Sun's outer layers. Raindrops split up sunlight into a band of about seven colours, to produce a rainbow; the light is refracted inside the water droplets, and broken down into the constituent colours. However, there isn't much useful information about the Sun to be learned from studying rainbows, because they are imperfect spectra.

The first serious solar spectroscopist was that giant of Cambridge astronomers, Sir Isaac Newton. Newton's scientific work ranged widely over mathematics, the behaviour of light, astronomy and gravitation. In the end he finished up as a mildly disturbed genius charged with running the Royal Mint. Newton's link with monetary bureaucracy is commemorated on an English one-pound banknote issued in 1978. This summarizes on its reverse side, and in an erroneous way, some of Newton's scientific achievements, including planetary theory, spectroscopy, and the invention of the reflecting telescope. Newton did many experiments in optics and showed that white light is split into colours by a prism. In 1665 he tried spreading out sunlight by means of a glass prism. He used a narrow slit of light emerging from window shutters. Another discovery came when W. Herschel, an experimental scientist, placed a thermometer bulb in the various coloured bands to see what temperature would be registered. The readings increased from blue to red. He then found that when the bulb of the thermometer was moved beyond the red light and into the invisible part of the spectrum the temperature did not fall, as he expected, but actually rose further still! Herschel had stumbled across invisible heat radiation and thereby started a new branch of science: infrared astronomy.

In order to examine the solar spectrum further, William Wollaston, who was partially blind, passed the light through a narrow slit and then through the prism. He found, in 1802, that the coloured light is chopped up by dark bands. He counted seven: two in the red part, three in the green, and two in blue-violet. This remarkable discovery, of apparent gaps in the spectrum of sunlight, led many other scientists to watch the Sun. Joseph von Fraunhofer of Munich produced pure spectra in 1814. He went on to map out 500 of the lines, an

The 50-metre solar tower at the Mount Wilson Observatory, near Los Angeles. (Hale Observatories, California)

Big Bear Observatory is a solar telescope in a lake. The body of water helps to steady the air during the day. (Big Bear Solar Observatory, California)

achievement now remembered by naming all the dark lines in the Sun the *Fraunhofer Lines*.

In 1821 Fraunhofer began improvements to solar spectroscopy by using a grating instead of a prism. The grating works on the principle of diffraction, whereas a prism exploits the refraction, or bending, of light. The latter occurs because light travels more slowly in dense materials in a manner that depends on wavelength: red light travels the slowest. Diffraction is a little harder to understand, but the basic phenomenon is as follows.

When a ray of light encounters the edge of a solid barrier it *slightly* loses its sense of direction, and a diverging beam of light scatters from the edge. If there are many 'edges', as in passing through a 'grating' of hundreds of fine wires, then the incoming ray of light is scattered, or diffracted, into a series of expanding beams. In certain directions of view all the light of a particular wavelength emerging from the grating is seen to be reinforced, or added together, if the difference in distance from one grating element to the next is a whole wavelength so that, as the angle of view across a grating changes, different wavelengths are seen intensely. In other words, a spectrum is made. If this sounds baffling, and it is if you haven't ever seen it done, try the following experiment. Make a shallow pool of water in your bathtub and place two solid barriers across in order to make a narrow (1-cm) gap in the water, like a toy harbour. Now make a series of parallel waves by waving a ruler gently and regularly up and down just outside the mouth of the 'harbour'. Inside the harbour you will see a series of semicircular waves travelling away from the mouth, not parallel waves. If there were many harbour mouths (a grating), the waves would strongly reinforce in some directions of motion and

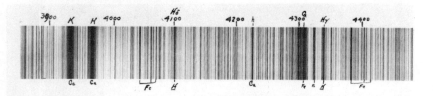

A small portion of the solar spectrum showing the numerous dark absorption lines first studied by Wollaston and Fraunhofer. (Hale Observatories, California)

cancel out in others. Just this effect occurs in a grating. One final example: you may have noticed a rainbow effect if you hold a record disc at a certain angle to the light. In this case the grooves of the record are acting as a grating as they reflect the light into your eye.

Large gratings have been used for solar spectroscopy, perhaps 10 centimetres (4 inches) square and ruled with tens of thousands of lines. By combining the telescope and spectrometer it becomes possible to examine the spectrum of different parts of the solar disc, and so investigate changes of temperature, composition and velocity in the outer layers of the Sun. Each and every layer of the solar atmosphere is characterized by a distinct range of temperature and pressure. Therefore each layer displays its own characteristics. Just as geologists can peel back layer after layer of the Earth's history, so spectroscopists can slice through the onion-like solar atmosphere.

Photographers sometimes use coloured filters to increase contrast. If they wish to photograph cloud formations they may put a red filter in front of the camera lens, and thus cut out the blue light from our clear sky. Similar techniques are valuable in astronomy: a spiral galaxy photographed mainly in blue light shows prominent arms; a gaseous nebula in red light displays all its delicate tracery. But these coloured glasses or gelatin filters are *broad-band*, that is to say, they let through light across a broad spread of wavelength, perhaps as much as 100 nanometres (a nanometre is one-billionth of a metre, or is sometimes still expressed as 10 ångstroms, i.e., one hundred-millionths of a metre). For scientific work much finer filters are available, and for solar photography they can be as narrow as 0.01 nanometres, or one-tenth of an ångstrom. These very narrow bandwidth filters are made on an interference principle: light is made to reflect inside the filter in such a way that all but the desired wavelength is cancelled out to a precision as small as 0.01 nm. (Only filters constructed on this interference principle should be used for direct visual observation, as discussed on p. 35.) An interference filter can cut out as much as 99.95 per cent of the incident light except at one favoured wavelength where all is transmitted. The beauty of these filters is that they allow the solar disc to be examined at very precise wavelengths. As we shall see, this gives astronomers an invaluable method for peering at different layers in the Sun's atmosphere.

Sunlight emerges from a variety of layers in the solar atmosphere. As I have hinted already, the temperature and pressure vary within the outer layers of the Sun. Red light emerges at a somewhat lower level than blue light. The yellowy-white light that we see by the naked eye is a mixture of light emerging from a range of heights. By peering at the Sun in specific colours, that is to say particular wavelengths, we look at particular layers of the solar onion. This is a very powerful tool for taking apart the solar atmosphere. The advantage is much increased when observations are made in one of the Fraunhofer spectral lines. To give one example: photographs, termed spectroheliograms, taken in the light of ionized calcium atoms (the K line at 393.4 nm) show bright regions where the calcium atoms are unusually excited, particularly near to sunspots.

Credit for the invention of a device capable of making a solar photograph in a narrow wavelength range, that is in monochromatic (one colour) light, goes jointly to George E. Hale of the USA and H. Deslandres of France. These two invented the spectroheliograph at the same time and each independently. Hale, a genius of American astronomy, made the first instrument for his private observatory located near Chicago. He was an undergraduate at the Massachusetts Institute of Technology when, in 1889, at the age of twenty-one, he modified a Harvard spectrograph so that it would image the Sun in a single spectral line. The basic principle of the method is fairly easy to understand. A solar telescope forms an image of the Sun on the slit of

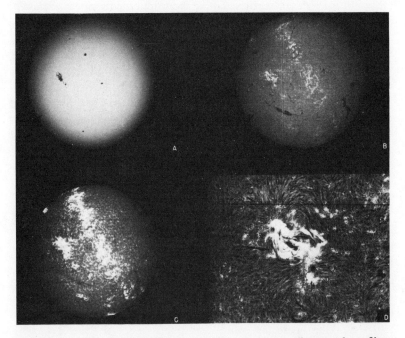

Three views of the Sun in different wavelength regions illustrate how filters enable astronomers to plot the structure in different layers of the atmosphere: A, white light; B, hydrogen light only; C, calcium light only. The enlargement, D, shows a flare region, captured in the red light of hydrogen. (Hale Observatories, California)

a spectrograph, so only one narrow strip of the Sun enters the spectrograph. The prisms, or grating, break this strip into the spectral lines, which are essentially monochromatic images of the spectrograph slit. By suitably positioning a photographic plate everything can be so arranged that only one strong line, say that of hydrogen alpha, falls on the plate. Thus the monochromatic image of one narrow strip of the Sun is recorded. If the image of the Sun's disc is now moved across the spectrograph slit, and at the same time the photographic plate is moved, a continuous image of the Sun is drawn out much as a raster scan builds up a television picture line-by-line. That's the idea behind the spectrohelioscope. Today it is much easier to employ the interference filters to do this job, and they are sold at prices which need not daunt the keen amateur observer. These filters have no moving parts and great advantages in terms of speed and uniformity of operation.

Hale's work on the solar spectrum led to the invention of another instrument for solar astronomy, the magnetograph. In June 1908 Hale noticed on a high-resolution spectrum that the spectral lines of the radiation from sunspots were split into several lines. In 1896 the Dutch physicist Pieter Zeeman had shown that if you put atoms of a light source into an intense magnetic field you break down certain of their spectral lines into further components. This happens because the energy levels of the outer electrons in the atoms get disturbed into a series of sub-levels when strong magnetism is around. Under this circumstance individual spectral lines become blurred or split into two or more components.

Astronomers who were aware of Zeeman's work suspected that the spectral lines in sunspots were blurred. To prove that magnetic splitting was taking place it would be essential to use a telescope of high resolving power on a good site. In 1905 Hale started a project at Mount Wilson, California, which took three years to provide unequivocal evidence of strong fields in sunspots. Hale's exciting discovery showed that sunspots have giant magnetic fields, for the sunspot lines were each divided into several lines, just as Zeeman had observed in the laboratory. The light from a particular sub-level of energy is polarized. This means that with a suitable arrangement of polarized filters it's possible to isolate polarized lines, caused by magnetism, from the solar radiation in a way that allows the magnetism itself to be mapped. This is now carried out on a routine daily basis at

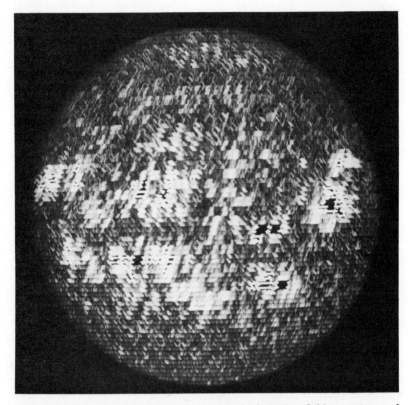

This magnetic map of the Sun's disc shows the location, field intensity, and direction of weak magnetic fields. The data are obtained by an instrument that exploits the Zeeman effect. The calibration strip shows how the recording line varies in appearance for different levels of magnetic field. (Hale Observatories, California)

the major solar observatories. The exploitation of the polarization properties enabled H. W. and H. D. Babcock to develop a very sensitive magnetograph at Mount Wilson in 1952. Scientists can use the data on solar magnetism to identify areas of major disturbance just beneath the visible surface. Apparently, if the magnetic field close to the surface gets sufficiently irritated by local disturbances it can be increased in strength until the field lines break through into the visible region. The magnetograph records these eruptions. Additionally it keeps track of the Sun's steady magnetic field, which reverses its polarity every eleven years or so. It is possible to use the magnetograph to see magnetic fields at various levels in the solar atmosphere, including prominences and the low corona.

The corona is the Sun's outermost region. You cannot see the very thin gas in this region except during a total eclipse of the Sun. This is because the relative intensity of the light even from the inner corona is down by over a million times compared to the intensity of the brilliant disc. Furthermore, the flood of light from the disc is scattered in Earth's atmosphere making the daytime sky glow with a pure blue light that is at least as bright as the pure coronal light. During a total eclipse, the brilliant disc is of course blanked off by the Moon and the sky goes much darker. Then the crowning glory of our daytime star briefly flashes into view: a halo of fire surrounding the black Moon disc.

For astronomers wishing to study the corona, total eclipses are inconveniently rare and last only a few minutes. Add to this the fact that many of them occur in remote parts of the world, or worse still in cloudy regions, and you have a powerful incentive for developing an artificial eclipse maker, which is exactly what the French astronomer B. Lyot did in the year 1931. His ingenious device, the coronagraph, lets solar scientists examine the faint light of the outer corona at almost any time. Lyot did not use any fundamentally new principle; he set out along a path that many had tried before. In his case a judicious mix of patience, attention to detail, and luck enabled him to succeed.

Essentially the coronagraph consists of two telescopes in series. The first telescope forms an image of the Sun, which is artificially eclipsed by a metal disc. The second telescope then throws this eclipsed image onto a photographic plate or film. It sounds simple enough; to make it work, however, much thought has to be given to eliminating the scatter of light. Stray light bouncing around inside the coronagraph would soon fog the plates. The objective lens is therefore made of a single piece of flawless glass—no bubbles, scratches or fingerprints must be present or they would scatter the light. The instrument has baffles and diaphragms to remove spurious light. It must then be operated at a high altitude in a transparent atmosphere.

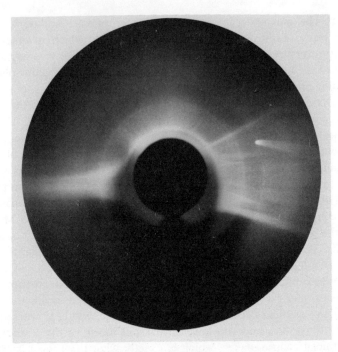

A coronograph was included in the Skylab instrument package. This photograph records coronal streamers and Comet Kohoutek the day before its closest approach to the Sun. (High Altitude Observatory, Boulder, Colorado)

Lyot operated his first coronagraph at the Pic du Midi observatory, 2,868 metres (9,410 feet) up in the Pyrenees.

When used in conjunction with a spectroscope or filters the coronagraph can be used to study the behaviour of particular atoms or elements in the corona. If it is combined with a spectrograph then it will give information on physical conditions, such as temperature or velocity, within the corona. Lyot was able to discover many new emission lines of highly-ionized atoms with his coronagraph. The coronagraph is the last of the optical instruments that I wish to describe. Now it is appropriate to mention briefly the techniques that are used to 'see' the optically invisible radiations from the nearest star.

Our Sun is an important source of radio waves from space. At most radio wavelengths it is the brightest object in the sky, although it is rivalled by the radiation from a supernova remnant (Taurus A, the Crab Nebula) and several radio galaxies (Cygnus A and Centaurus A, for example). In the optical domain the brilliant Sun has no competitors, but this is not so true in the radio spectrum. Here the Sun is challenged by distant but extraordinarily powerful objects. Most of the telescopes used for solar radio astronomy collect the waves by one or more metal bowls or meshes. These parabolic dishes reflect the radiation and bring it to a form where it can be detected and measured. At metre and decametre wavelengths arrays of aerials or antennae are often used instead of reflecting dishes.

Radio telescopes cannot perceive structure in the Sun's radio image unless they are suitably arranged and connected to form a radio interferometer. Radio waves are millions of times longer than light waves. This means that the resolution, the capacity to detect fine detail, is far worse in the case of radio telescopes. This can be overcome by pairing nearby telescopes to make interferometers. The separate reflecting elements need to be spaced by hundreds of wavelengths in order to get adequate resolving power for useful observations of the Sun. A couple of dishes separated by a kilometre, for example, and observing at a frequency of (say) 300 MHz (1 metre wavelength) can resolve detail with a scale size of 3 minutes of arc. At the surface of the Sun this angular resolution converts to a length of roughly 150,000 kilometres. Clearly, radio astronomers cannot expect to map the radio Sun with anything like the detail seen in the finest optical photographs.

Perhaps the best-known solar radio telescope is the Australian radioheliograph at Culgoora in New South Wales. This instrument is an array of ninety-six dishes, uniformly arranged in a circle with a diameter of 3 kilometres. Under computer control the signals from all these sun-trackers are combined to produce an image of the radio Sun twice a second. At that rate it's possible to use a series of radio pictures and make a movie of the activity of the radio Sun.

A device known as the radio spectrograph is immensely important. Used in conjunction with a radio telescope, this instrument records the radio spectrum over a given range of frequencies as a function of time. The instrument finds important applications in the study of sudden bursts of radio emission.

The application of space-age technology to sun-watching has forged the greatest revolution in solar astronomy since Galileo's telescope. Ultraviolet and X-ray emissions from space cannot travel through the Earth's atmosphere. These radiations are particularly important because they originate in the active zones of the Sun's outer atmosphere. Their message tells of the Sun in turmoil, of physics pushed to the limits, of matter and energy in fearsome conflict. The high-energy radiation can be detected only by instruments taken above the atmospheric blanket by balloons, rockets and satellites. Astronomers went up in balloons way back in the eighteenth century and used planes in the early twentieth century. In the pioneering days of X-ray astronomy, rockets were used exclusively. Far more data are obtained, however, from a satellite in permanent orbit round the Earth or the Sun. In the case of studying the Sun there is also the unique and exciting possibility of actually sending non-orbiting probes close to the surface of a star. Satellite instruments may also allow the Sun to be continuously watched, uninterrupted by clouds or nightfall.

A few elements of the solar radio telescope at Culgoora, New South Wales. (CSIRO Division of Radiophysics, Epping, New South Wales)

Rocket-borne instruments obtained the first X-ray images of the Sun. These showed, for the first time, highly-disturbed regions of X-ray emission in the corona.

Several series of space probes have expanded our knowledge of the Sun. For example, the IMP spacecraft (Interplanetary Monitoring Platform) made hundreds of measurements of the electrons ejected from the Sun in the late 1960s. Soviet scientists have used the Prognoz satellites for making X-ray and gamma-ray measurements of the active Sun. The scientific equipment deployed in Prognoz 2 (launched 1972), for example, included spectrometers for X-rays and gamma radiation, and detectors for atomic particles (electrons, protons and neutrons) expelled by the Sun.

The eight Orbiting Solar Observatories (OSO) carried a battery of instruments. OSO 7, which was launched late in 1971, functioned for two and a half years, a remarkable achievement for a satellite that seemed doomed soon after launch as it went into an apparently uncontrollable spin. Only a new Sun sensor and gyro, not used on any previous OSO craft, enabled the ground controllers to establish their authority. On board OSO 7 was a white-light coronagraph, which had its occulting disc (artificial moon) simply extended on a rod in front of the telescope; in the vacuum of space you only need to hold a small disc at arm's length to make an artificial eclipse and see the corona! Other instruments included a gamma-ray spectrometer, a spectrograph for X-rays and high-energy ultraviolet radiation, and a solar X-ray instrument. Other astronomers got a look in too, because OSO 7 carried celestial X-ray telescopes for peering at other stars. As a result of drag every time its 90-minute elliptical orbit cut through the top of Earth's atmosphere, this magnificent spacecraft finally crashed back through the lower atmosphere in July 1974. Between them, the OSO satellites managed to watch the Sun throughout a full eleven-year cycle of solar activity.

The most phenomenal success in the 1970s was the Apollo Telescope Mount (ATM) solar experiment carried aboard the American space station Skylab. This was the first real manned astronomical observatory in space. At the tail end of the lunar programme there came an opportunity for solar astronomers to collaborate on a concerted attack by putting a battery of eight solar telescopes into space and controlling them from the ground and with the aid of resident astronauts. The six principal solar instruments/telescopes recorded and photographed the outer atmosphere of the Sun at wavelengths ranging from the visible light through to X-rays. The X-ray pictures showed the Sun in exquisite detail out to a height of half a solar radius above the surface.

Many significant advances in instrumentation were tried on Skylab. For example, the Skylab's size enabled large and massive instruments to be carried. The experimental area was 3 metres (10 feet) long by 2 metres in diameter, and this housed almost a ton of telescopes. The system could call on 2 kW of electrical energy to control the experiments, an improvement of 100 over the OSO satellites. These factors meant that the instruments were significantly improved in their performance, in some cases by hundreds or thousands of times.

Just as important also there was plenty of provision for sending data back to Earth at a fast rate—600 photographs per day on average. Furthermore, several views of the Sun could be displayed simultaneously—white-light and ultraviolet for example—to let scientists and astronauts plan their observations for maximum effect. The fact that Skylab was manned meant (obviously) that the crews returned to Earth. This opened up a marvellous opportunity to use film in conjunction with several of the telescopes. Even within the electronic wonderworld of spacecraft, film has great attractions as a data-storage medium. One image of the Sun may be split into a million pieces that have to be transmitted, TV-style, to the ground. Film provides a means of fast, efficient storage without the need for elaborate digitizing and transmitting. The presence of astronauts meant that film could be retrieved (this meant leaving the interior of Skylab in order to remove film magazines) and eventually brought to Earth. In all, the scientists exposed thirty canisters of film on which there were 150,000 successful exposures.

Another factor that helped the Skylab mission was ground support. A world-wide network of solar observatories and stations kept an independent watch on the Sun in order to assist the Skylab researchers. When something unusual happened they could aim the eight solar telescopes from space. Every day the mission controllers worked out an observing programme for the following day. And, finally, the astronauts played a crucial role in patching up the spacestation and the equipment as necessary—by freeing jammed shutters, replacing cameras, retrieving film and even salvaging the space capsule itself at the start of the mission after a catastrophic launch. The cost of all this, about $250 million for the solar experiments, was well rewarded by the large amount of new information it yielded on the performance of the astronomers' favourite star.

One last point is worth making: with the exception of the radio telescopes, many of the instruments used in solar research cannot be used to investigate other stars, which are so very much more distant than the Sun. The nearest stars are something like a million times as far away. We cannot see the faint haloes, revealed for us during solar eclipses, round other stars. It is only possible to make out very sketchily even the crudest surface details for a handful of nearby giant stars. Not even the largest celestial telescopes can resolve any disc for the stars of solar size. No other star can be glimpsed with a precision anything like that routinely achieved by solar astronomers. The daytime star, full of mystery, tells us more about ordinary run-of-the-universe objects than any single night-time star ever will.

The Architecture
of the Sun

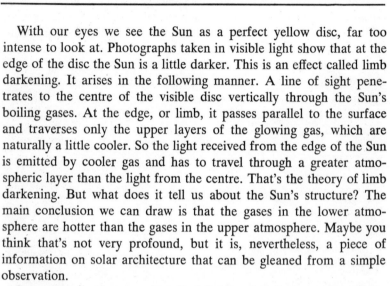

With our eyes we see the Sun as a perfect yellow disc, far too intense to look at. Photographs taken in visible light show that at the edge of the disc the Sun is a little darker. This is an effect called limb darkening. It arises in the following manner. A line of sight penetrates to the centre of the visible disc vertically through the Sun's boiling gases. At the edge, or limb, it passes parallel to the surface and traverses only the upper layers of the glowing gas, which are naturally a little cooler. So the light received from the edge of the Sun is emitted by cooler gas and has to travel through a greater atmospheric layer than the light from the centre. That's the theory of limb darkening. But what does it tell us about the Sun's structure? The main conclusion we can draw is that the gases in the lower atmosphere are hotter than the gases in the upper atmosphere. Maybe you think that's not very profound, but it is, nevertheless, a piece of information on solar architecture that can be gleaned from a simple observation.

Some stars, by the way, show limb brightening (hotter at the edge), which means that the run of temperature with altitude is opposite to the Sun. And, just to complicate the picture, radio maps of our Sun show more intensity at the edge of the disc, indicating that some of the radio emission is generated in the outer atmosphere.

To explore the Sun's structure I am going to describe an imaginary voyage from the centre of the Sun to the Earth—imaginary only for matter (humans!), because packages of radiation, called photons, are making this journey all the time, bringing light and heat to the world.

As we go on the journey you might wonder *how* we know a temperature or density. The values of most of the physical parameters are not measured but instead are derived theoretically. The structure of the interior has to be deduced by pure reason—theory, equations, luck and powerful computers. All that is known are certain global properties, such as the mass and the radius, and the physical conditions at the radiating surface. By means of observations of other stars we also know how some of the properties (for example, surface temperatures) depend on other properties (such as the mass). The chemical make-up of the Sun is also very well documented, as we shall see, from spectroscopy. What the theorist has to do is to take all these data and then deduce in a self-consistent manner how the Sun works. This guesswork is used to make a mathematical model of the Sun. If the model fits all the observed properties, and continues to do so when new data come in, then it's probably a fairly good approximation to the real thing. This guesswork has been going on for half a

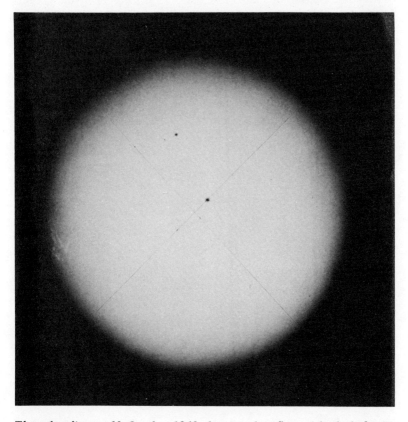

The solar disc on 12 October 1963 showing the effect of limb darkening and the presence of two sunspot groups. (Royal Astronomical Society, London)

century now. We have a reasonable global picture of what's going on in there. In fact the stage has been reached now in which one cannot fiddle around too much with properties such as the central temperature without affecting seriously the observed brightness of the Sun. So, to sum up, our imaginary journey starts deep inside the Sun, a region that so far can only be explored by mathematics and computer science.

The central region of the Sun is called the core, which is really only a shorter word for central region. Inside the core, matter is enormously compressed by the almost unbearable weight of all the stuff beyond and outside the core. The Sun is held together by the gravitational attraction of its own matter, and the core is crushed by the weight of the overlying material. Although the core only extends to a quarter of a solar radius, and therefore accounts for under 2 per cent

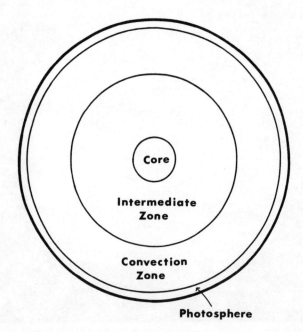

The main zones into which the solar interior is divided. The dense core contains half the Sun's mass and it is here that the nuclear energy is released.

of the total solar volume, about half the Sun's mass is packed into it. Packed is the operative word—the central density is 155 grams per cubic centimetre, about a dozen times as dense as lead. The internal pressure is a horrendous 3,000,000 million (3×10^{11}) atmospheres, and the temperature 14–15 million degrees Kelvin.

These conditions are precisely those needed to operate a nuclear furnace, for that is what the core is: a controlled nuclear power station in which hydrogen is converted into helium. The energy released by the nuclear processes travels through the core as radiation.

About a quarter of the way from the centre we leave the fiery core and cross into the convection zone, which extends right to the visible surface. The other half of the Sun's mass, the half which is not in the core, is in this zone. No energy is created here because the temperature and pressure of the material fall below the values needed to sustain the nuclear firecracker. All the way to the surface the temperature and pressure fall steadily. We are progressing down a *temperature-pressure gradient*. One-tenth of a solar radius beneath the surface, for example, the temperature is about 600,000 K and the pressure only 1 million atmospheres. Within the convection zone large-scale movement of material is present, a seething mass is boiling away, carrying energy from the core out to the surface.

At the visible surface, which astronomers call the photosphere, it becomes possible to 'see' for a reasonable distance. The inside of the Sun is very opaque (if it were transparent we would be able to see right through it); therefore our imaginary traveller could only see about one centimetre in any direction deep inside it. The photosphere is the transition layer where the material has cooled enough to become transparent. Light can leave this surface more or less unhindered, and the corollary is that we really can see the surface. Another important fact to note here is that the yellow disc of the Sun has a very sharp edge, not the fuzzy appearance that we might expect for a ball of luminous gas. This sharpness is due to the very sudden transition from high opacity to high transparency. The white light that we see comes almost entirely from the layer in which this crossover of properties takes place. This layer is about 500 kilometres thick, less than 0.1 per cent of the Sun's radius—that's why the edge of the visible Sun is so sharply defined. Now we can understand a little more clearly the phenomenon of limb darkening: a line of sight to the centre of the disc terminates about 500 kilometres deeper, and therefore reaches hotter and brighter layers, than a sightline to the limb.

At the surface our thermometer has dropped to around 6,000 K, the pressure to only one-sixth of an atmosphere, and the density to a trivial amount—less than a millionth of the density of ordinary water.

Now we move further out through the onion-skin layers of the Sun's outer atmosphere. Above the yellow-white photosphere lies the chromosphere, a cool region that is seen to flash into view with a pinkish light for a few seconds during a total eclipse. No hard and

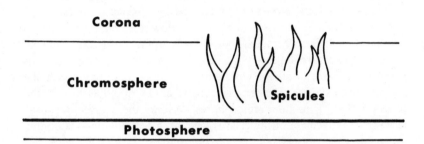

The Sun's atmosphere is conventionally considered as a series of layers. In practice these merge imperceptibly and their actual extents vary with the degree of activity on the surface.

fast boundary exists between the photosphere and chromosphere. We know we've reached the latter when the temperature reading falls to about 4,500 K. From then on it *rises* as we continue outward, reaching 10,000 K in the upper chromosphere, thereafter rising sharply to register 1 million degrees at the top, a few thousand kilometres above the photosphere. The density, meanwhile, has dropped to 10^{-16} grams per cubic centimetre (about ten million hydrogen atoms per cubic centimetre).

The final layer of the atmosphere is the corona, which extends out to at least ten solar radii. Within the corona the temperature is consistently 1 million degrees or more. Material in the corona is almost entirely transparent to visible light, so its own glow is very feeble. That's why it can only be seen during total eclipses.

The corona is a strong source of X-rays. When atoms are heated up to a temperature of a million degrees or so, only the heavy atoms, such as iron, are able to hold on to any of their orbiting electrons, and even they can only keep one or two in grasp. These heavy atoms, stripped down to just a couple of electrons, produce emission lines at X-ray energies. Atoms that are able to recapture an electron also send out X-ray photons. And there is another effect: the violent jiggling around of atoms at the high temperatures in the corona produces X-rays as atoms stripped of electrons brush close to each other.

Beyond the corona our imaginary travellers come to the solar wind. This is generated within the corona. In effect the very top of the corona, millions of kilometres above the Sun's surface, is breezing away into space. The Sun's gravitational pull is not strong enough there to keep a hold over the corona. Instead of being in equilibrium it is evaporating into space as a wind of particles. The wind starts with a speed of about 4,000 km per second. By the time it reaches Earth this interplanetary gale has blown out, and the breeze drifts past our planet at 400 km per second—about a million miles an hour. Mind you, you would not be knocked over by the high-speed solar wind because the density of matter in the wind is small, about 1,000 atoms in a teacup-sized volume of space. In the course of a year the Sun loses 200 million million tons (3 million tons a second) of matter by this means; the exact amount varies because it is influenced by the state of activity on the Sun.

The discovery of the solar wind preceded the launch of satellites such as IMP. Surprisingly, the finding of the solar wind was a result of astronomical observations of extremely remote radio sources, billions of light years from the solar system, known as quasars. In 1964 radio astronomers at Cambridge found that as the Sun approached the line of sight to a distant radio source, a fluttering in the radio signal received occurred. This effect, grandly called *interplanetary scintillation*, happens in much the same manner as the twinkling of stars in the night sky. Irregularities in the solar wind—squalls and calms in the particle breeze—distort the path of radio waves and cause the source to 'twinkle'.

Indirectly, the discovery of the solar wind led to yet another unanticipated result. The Cambridge researchers built a special telescope for monitoring the solar wind and for watching the influence of the wind on radio sources. Within a few months of operation this new instrument found the first pulsars. Pulsars are fast-rotating neutron stars, balls of nuclear matter about 10 kilometres in diameter with a mass similar to the Sun. Theorists had predicted their existence for decades, but nobody had worked out how to find them in the cold reaches of interstellar space. The Sun, quite accidentally, showed astronomers the right way to go about it!

The planets moving on their elliptical orbits round the Sun are really cruising within the Sun's outermost layers. The two planets with strong magnetic fields, namely the Earth and Jupiter, deflect the main onslaught of the solar wind by means of a magnetic cage, termed the magnetosphere. Our imaginary travellers would detect changes in the magnetic environment in the vicinity of the Earth. A shock wave in the wind is located immediately in front of the magnetic buffer. Once in this vicinity, of course, the trip from the Sun's heart to the frozen outposts of the solar system is not so speculative since men have actually been through this region on their way to the Moon. Additionally, the Earth's magnetic shroud has been probed by many spacecraft.

By the time we pass the orbit of Saturn, and pause for a moment at a distance of 1 billion miles from the Sun, we have come to the place where the Sun essentially peters out. Now it is indistinguishable from the interplanetary medium, the wisps of gas and specks of dust that wander through the realms of the planets. This dust is responsible for a beautiful Sun-related phenomenon, the *zodiacal light*. Also known as the false dawn, it is a cone of light in the sky, visible to the naked eye over to the west just after sunset or, in the east, just before the sunrise. It is, in fact, sunlight being scattered from the dust in interplanetary space. On a dark moonless night it contributes about one-third of the total light in the sky. I have never seen it convincingly in England, but was most impressed by its prominence in Australia. It is also visible frequently from dark sites in the southern United States.

Once we leave our solar system the Sun is truly like most other stars. It is type G2, which is very common. But, as this fictional journey has shown, the Sun is the only star that we can take to pieces. Although we can picture the interiors of other stars, we cannot resolve their surfaces, map their coronas, detect weak stellar winds, or see the day-to-day changes in their atmospheres. Of course there are uncertainties in the picture sketched above; the crucial point is that the blurring of the picture is very much worse when we look at the other faraway stars.

The dim zodiacal light is sunlight reflected from fine dust scattered throughout the planetary system. This photograph was taken on the Apollo 17 mission and shows the cone of light against a starry background. (NASA and National Space Science Data Center, World Data Center A, Greenbelt, Maryland)

The Alchemist's Crucible

A star is a battleground with two forces held in balance. All stars are globes of gas held together by the force of gravity. How do we know this? The answer is that the outer layers or atmospheres of stars, including the Sun, are quite definitely composed of gas, and the inaccessible inner regions are so hot that they could not be anything but gas, or to be more exact, a plasma.

Now, every particle within the Sun is pulled this way and that by the gravitational force exerted by all the other particles. Gravity is a force that, unlike magnetism, always pulls, never pushes; so the Sun is being squeezed all the time by its own gravity. This shrink-wrapping of the gravitational force prevents the Sun from merely wafting away into space. A little calculation soon shows that it is held together very effectively. At the surface temperature of 6,000 degrees the individual atoms are buzzing along at speeds of nearly 10 km per second. If gravity were switched off they would travel out to a solar radius above the surface in only a day. So within just a day the Sun would double in size—in short it would explode into space and dissipate itself within a few weeks if gravity suddenly became ineffective. Without gravitation there would be no stars, no Sun, no us! However, this inward force has to be counterbalanced. In the case of the Sun, all the material would plunge inwards at a catastrophic speed were gravitation the *only* force. The free-fall time for the Sun is about half an hour, so what is the magic force that stops it caving in?

The force is the internal pressure of gas. When you blow up a balloon, gas pressure stretches out the elastic skin. Gas pressure also holds up the Earth's atmosphere, which would otherwise just fall down from the skies. Strictly speaking, it is the gradient of the pressure that supports the overlying regions within a star like our Sun. The observation that stars, particularly the Sun, are exceedingly stable over millions of years, permits us to conclude that every little part of the Sun is held in a perfect balance between gravity pulling in

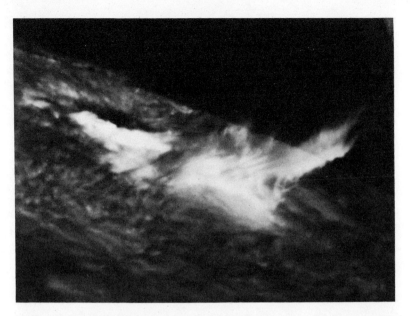

Explosive energy release in the photosphere sends a spray of matter surging away from the Sun. Dramatic photographs such as this give the impression that the solar surface is a raging inferno. In fact the energy that is bursting forth so furiously here was originally manufactured deep in the solar interior. (Sacramento Peak Observatory, AFCRL, Sunspot, New Mexico)

and the pressure gradient pushing out. The pressure at the centre of the Sun, billions of atmospheres, can be roughly estimated simply by calculating the pressure needed to resist the 'weight' (gravitational force) exerted by a column of gas stretching all the way from the centre to the photosphere.

The gas laws of physics show that the pressure and temperature of a fixed amount of gas are related. Given the central pressure (2.5×10^{11} atmospheres) the calculation of the central temperature inevitably follows from the gas laws and is seen to be in the region of 10 million degrees. Accurate calculations give 14–16 million degrees.

Now that we have established that the Sun is a fiery ball of gas held in perfect equilibrium, we must answer an important question. Since the hot surface radiates energy at the rate of 3.83×10^{26} watts, how does it compensate for the relentless energy loss? Without compensation the Sun would inescapably cool, shrink, and finally collapse.

However, we know that the Sun has in fact maintained essentially the same energy flow for about 5 billion years.

Not so very long ago astronomers had not really got the vaguest idea as to what kept the Sun going. I have a book published in 1892 in which the solar energy is simply explained away thus: '[The Sun] is a mighty furnace of heat and flame, beyond anything we can possibly imagine.' The same book states that astronomers still consider that the Sun '. . . may have a solid and even cool body within the blazing covering.'

One of the first serious attempts at a theory dates from 1842. In that year. J. R. Mayer suggested that the Sun's energy became replenished by infalling meteorites. It did not take long to show that this could not possibly sustain our Sun for long.

As an alternative, Lord Kelvin (William Thomson) and Hermann von Helmholtz tried a gravitational theory. They worked out that if the Sun contracted annually by a mere 20 metres, enough heat energy would be released by this extra compression to account for the measured luminosity. This slow contraction, far too small to be measurable even today, would keep the Sun shining for about 50 million years. That seemed to be just long enough to explain everything, given the hazy state of knowledge of the geological timescales in the mid-nineteenth century. But by the early twentieth century it was abundantly clear that the fossil and sedimentary record stretched back for hundreds of millions of years and that something far more effective than mere burning or shrinking was needed to keep the Sun alight. (Only since the manned missions to the Moon has it become certain that the Earth and Moon are about 4.7 billion years old, and the Sun about 5 billion years old.)

The astronomers were therefore stuck with a Sun that could not perform for as long as the geologists said it should. Very embarrassing. To the rescue came Sir James Jeans, who in the 1920s developed the idea that radioactivity would provide the vital heat source. Although this hypothesis is now known to be incorrect it set other scientists thinking along the right lines: namely, the transformations in the atomic nuclei must be the energy source for the Sun, and hence for other stars too. Radioactivity is a process in which the nuclei of atoms decay, generally by emitting electrons, with a small release of energy. Jeans guessed that super-atoms, which he thought could have been left over from the early history of the Universe, might be the energy source that fuelled the Sun.

The Jeans hypothesis eventually proved to be inadequate, however, and it is instructive to see why. The problem posed by the hypothesis is that if the Sun, and other stars, are fuelled by radioactive decay, why are they so stable? We have already glimpsed the tight equilibrium inside the Sun: gravitational and pressure forces in perfect balance. This is impossible to organize with radioactive heating because radioactive decay is a *spontaneous* process seated in the nucleus itself which is therefore not sensitive to external conditions such as temperature and pressure. In short, radioactive stars, if they could exist, would have no sure means of controlling the energy release and temperature. Such stars would either fizzle out and implode or explode like atomic bombs. What is needed, then, is an energy source that helps the star remain stable. We want an energy source that works harder if the star tries to collapse a little, thus providing more heat and a higher resisting pressure; the same source should slacken off if the star tries to expand, thus reducing the pressure. An energy source must also be a safety valve if a star like our Sun is to be stable.

In 1931 the British astronomer R. d'E. Atkinson suggested that the source of Sun-energy might arise from the capturing of extra protons by the nuclei of atoms. The proton is a heavy nuclear particle which carries a positive electric charge. The hydrogen atom consists of one massive central proton and one orbiting electron. This combination of proton and electron makes the atom electrically neutral.

It is at this point in our story that we can see an example of how the study of the Sun particularly has assisted physics generally in the last half century. Following the idea of Atkinson, the American astronomer Donald H. Menzel pointed out in 1932 that the high proportion of protons (hydrogen nuclei) in stellar interiors made it likely that the interaction of protons would be of great importance in solar energy. The union of protons to form heavier nuclei is accompanied by the release of nuclear energy. In the early 1930s, however, scientists believed that heat-releasing interactions between protons (so-called thermonuclear reactions) could not take place because the positive electrical charges on protons caused them to repel each other too strongly. George Gamow demonstrated that this simple argument was unsound in the strange world of atomic particles. Using the new science of quantum mechanics, the branch of physics which shows how these particles of the micro-cosmos interact, Gamow proved that protons could tunnel into each other. In a sense they could get near enough together for the nuclear glue to make them stick before the electrical force realized what was going on!

In 1939 Hans Bethe in the United States and Carl von Weizsäcker in Germany capitalized on the work of Gamow. They independently presented the first plausible pictures for the release of nuclear energy in the Sun. The formulation was obviously based on the knowledge of nuclear physics then existing and it involved sticking extra protons into the nucleus of carbon. We now know that reactions involving carbon nuclei are only important in stars more massive than our Sun. At the home base a simpler reaction involving just protons, essentially, is thought to be the important process. But before examining this miracle of nature, one more piece of background information is necessary, the most famous equation in physics.

We need to make a brief excursion into the baffling world of Einstein's Theory of Relativity, which contains among many other gems the relation $E = mc^2$. What this equation tells us is that energy (represented by E) and mass (represented by m) are interchangeable quantities. The symbol c^2 stands for the square of the speed of light, which is a very large number. To give a quantitative example: 1 gram of matter is energetically equivalent to 30 million kilowatt hours, enough electrical energy to keep an average home going for 10,000 years at the present annual rates of consumption. This example assumes that the mass (m) can be switched over to energy (E) with complete efficiency. Although complete switching may take place for certain fundamental particles it does not happen with high efficiency for bulk matter. As we shall see, the Sun is nothing like 100 per cent efficient.

The dominant energy-release mechanism in the Sun is the combining of four protons—hydrogen nuclei—to make one atom of helium. In this thermonuclear reaction 0.7 per cent of the proton mass is destroyed, because the resulting helium nucleus and a few other particles are not quite as heavy as the four protons consumed in the reaction cycle. This mass difference, actually a mass loss, appears as radiation, that is to say, as energy.

At first reading, the series of reactions that convert hydrogen to helium seems amazingly unlikely. Yet the Sun is up there in the sky so we know it must all take place!

In the proton-proton chain of reactions the first event is for two protons to stick together tightly and form a nucleus of heavy hydrogen or *deuterium*. This involves kicking out a positively-charged electron (positron) from the nuclear duet, for the deuteron is actually a proton and neutron glued together, so one of the participating protons has to transform into a neutron. Two unlikely events precede the appearance of a deuteron inside the Sun. First of all, one of the protons has to be going about five times faster than the average in order to have enough energy to tunnel through the repulsive electrical field of the other proton. Inside the Sun the distribution of particle velocities is such that only one proton in a hundred million is zooming along at five times the average speed. Then, during a collision with a second proton lasting for a thousand-million-million-millionth (10^{-21}) of a second, one proton has to switch into a neutron. Unlikely? Yes, this is extremely unlikely for any given proton. Any particular proton can only expect to manage this circus trickery after tens of billions of years' practice! If the event were not unlikely the Sun would explode instantaneously. Yet, there are so many protons in the heart of the Sun that 3×10^{38} of them take part in this unlikely reaction every second. We see the combination a low probability for the event and an enormous number of particles leads to a substantial interaction rate.

As the proton pairs collide reluctantly they spawn deuterons. Unlike their parent particles these are eager to combine with something else. Within just a few seconds the typical deuteron grabs a spare proton. Thus united, the new combination, a nucleus of helium-3 (^{3}He) has three possible fates. Any particular helium-3 nucleus cannot influence which of the three possible routes it will take, for the careers of nuclear particles are governed only by laws of chance. The most likely event (95 per cent chance) is that it teams up with a nucleus like itself. This pairing event leads to the final stage of the reaction chain: the formation of a nucleus of helium-4 (^{4}He, or the alpha particle; this is the nucleus of the common form of the helium atom) and of two protons.

Before examining the other two branches of the chain that fuels the Sun, let's summarize the main reaction. To do this I will write the reactions in a nuclear physics shorthand notation which can be easily understood. In these symbolic expressions ^{1}H denotes a proton, ^{2}H a deuteron, ^{3}He the light form of helium, ^{4}He the usual form of helium consisting of two protons and two neutrons. Three more symbols are needed: e^+ to represent the positron, or positively charged electron, ν to indicate the neutrino, a weird particle without charge or mass which very rarely interacts with matter, and finally γ for the photon, or electromagnetic energy (X-rays, light, radio waves, etc.).

The reaction described above is then written:

$$^1H + {}^1H \rightarrow {}^2H + e^+ + \nu$$
$$^2H + {}^1H \rightarrow {}^3He + \gamma$$
$$^3He + {}^3He \rightarrow {}^4He + {}^1H + {}^1H$$

The incoming particles are on the left of the arrow, and the products of the reactions on the right. The total effect of this branch of the proton-proton chain is to take six protons and process them to yield one helium-4 (^{4}He), two protons, one positron, one neutrino (see Chapter 6) and some energy.

What happens if we compare the mass of those six protons at commencement with the masses of the reaction products? In percentage terms, about 0.7 per cent of the mass of the four protons that ended up as the helium-4 nucleus is lost. Every kilogram of hydrogen processed in this manner loses 7 grams of mass which is converted to pure energy—6×10^{14} joules of it. To put the amount into context, it is equivalent to 200 million units of electrical energy from every kilogram of hydrogen. This rate of energy release vastly exceeds the performance of any chemical reaction and is also much greater than the heat release through the steady contraction of the Sun envisaged by Kelvin and Helmholtz. The sun's supply of hydrogen is so vast that it can keep going for some 10,000 million years at the present rate of emission, and is now about halfway through the reserve.

Now I want to return to the other two possible branches for the proton-proton chain. In the Sun the helium-3 particle, once formed, can instead (5 per cent chance) bump into a helium-4 and make beryllium-7 (^{7}Be) thus:

$$^1H + {}^1H \rightarrow {}^2H + e^+ + \nu$$
$$^2H + {}^1H \rightarrow {}^3He + \gamma$$
$$^3He + {}^4He \rightarrow {}^7Be + \gamma$$

This then branches one of two ways. Either the beryllium-7 catches an electron (e^-):

$$^7Be + e^- \rightarrow {^7Li} + \nu \quad (^7Li \text{ is lithium-7})$$
$$^7Li + {^1H} \rightarrow {^4He} + {^4He}$$

or it catches a proton and makes boron-8 (^8B), which, being unstable, immediately decays to beryllium-8 (^8Be) and thence to two particles of helium-4, thus:

$$^7Be + {^1H} \rightarrow {^8B} + \gamma$$
$$\downarrow$$
$$^8B \rightarrow {^8Be} + e^+ + \nu$$
$$\downarrow$$
$$^8Be \rightarrow {^4He} + {^4He}$$

This, then, symbolizes the three possible routes by which protons—the centres of hydrogen atoms—stick together and make ^4He—the nuclei of normal helium atoms. Only chance dictates which route the ^3He nuclei will take once formed, but the likelihood of each branch can be calculated. In the case of our Sun, almost all the energy (95 per cent) is in fact generated by means of the first of the three possible sequences.

It's worth pausing at this point and noting that the theory of thermonuclear energy generation (that is, making energy by destroying nuclear matter) was proposed and worked out for the Sun long before there was any possibility of conducting controlled thermonuclear experiments in laboratories. Only from the mid-1970s onwards has it been possible to create in the laboratory the temperature and pressure conditions at the centre of the Sun. This can be done fleetingly by focussing an extremely high-powered laser beam onto a droplet of heavy water, which gets compressed by about a million million atmospheres as it recoils in the searing beam. Heavy water, which is extractable from natural water, has the heavy hydrogen isotope, deuterium. This is used because the first stage of the proton-proton reaction is so unlikely that it has never been seen in the laboratory, although it goes on in the Sun. This again illustrates that the natural reaction demands the presence of enormous quantities of hydrogen, a situation that prevails in stellar interiors. But to get at this energy in the laboratory we must start with deuterium, not hydrogen.

The equations governing the release of nuclear energy in stars depend sensitively on the central temperature and pressure. We know how much energy is made at the centre because we see its effect at the solar surface. Every second the Sun eats its way through 655 million tonnes of hydrogen which is converted to about 650 million tonnes of helium. When our Sun came into existence, a little over 70 per cent of its mass was hydrogen. This central reserve is dwindling at the rate of 5 million tonnes per second. As a result of this consumption the Sun can only last out for a further 5,000 million years: our daytime star is therefore already middle-aged.

We are all too familiar with the fact that when nuclear fusion reactions are induced on the Earth the result is a gigantic explosion— the hydrogen bomb. If the Sun's nuclear magic could be tamed on a small scale then nuclear *fusion* power stations could be built, using hydrogen as a fuel. But, so far, this goal of nearly infinite supplies of energy has remained elusive. This raises the question: since the energy release is to fantastic why do stars exist at all? Why doesn't the Sun explode like a bomb right now?

Clearly the Sun is amazingly stable. After all, it hasn't changed a great deal in the last few billion years and it obviously is not a bomb. The Sun tames its nuclear fury in the following manner.

The nuclear processes are self-stabilizing in a reactor as large as a star. To see that this is so, imagine that some disturbance causes the Sun to expand a little. This imaginary expansion would lead to a fall in the central temperature and pressure. In that case the nuclear particles would no longer be dashing about so fast nor hitting so hard. Hence fewer of them would stick together. The release of nuclear energy therefore would take place at a somewhat reduced rate. This reduction would, in turn, lower the temperature and, most crucial of all, would diminish the outward pressure gradient. So the underlying push would automatically ease off if the Sun tried to expand. Similarly, any slight compression would raise the temperature and speed up the nuclear reactions just enough to resist the new pressure. Throughout most of its total life, and all of our human lifetimes, our Sun is in perfect balance, therefore, between the energy radiated and the production at the centre. Any short fall in energy production would be made good rapidly as the Sun shrank a little in order to pile on the pressure at the centre.

An important contributory factor to the stability of the Sun, one touched on earlier, is the great opacity of the solar material to radiation. Conditions in the core are such that a quantum of radiation (photon) only goes about a centimetre before colliding with a material particle. These frequent interactions make the radiation lose all sense of direction. The photons therefore wander aimlessly around until they chance to reach the photosphere. There the temperature is such that the matter quite suddenly becomes transparent to visible radiation, which then streams out at the speed of light to the Earth and the rest of the cold universe. It takes about 10 million years for solar energy to leak from the core to the photosphere. This long timescale helps to balance the Sun. If the Sun were imagined to be transparent the radiation would then stream straight out, without creating the stabilizing pressure gradient, and this would lead to a cosmic bomb.

The first step in solar energy release is the joining of two protons to make a deuteron. Under the conditions in our Sun this is an unlikely occurrence, but it does happen at the right rate to keep the Sun shining. If the stabilizing factors did not exist, an explosive situation would certainly develop. And, once the explosion had started, the temperature and pressure at the centre would rise rapidly. The increasing temperature would make the protons jiggle about faster and faster, so they would collide and stick most readily, and hence the energy-release rate would rise. In short, the bomb would go off. This does indeed happen towards the end of the lives of stars much more massive than our Sun.

We have now reached the stage at which I want to consider most carefully the ways in which energy actually moves through the Sun. In elementary physics three mechanisms for moving energy are discussed: conduction, convection and radiation. The first only occurs in solids so it is irrelevant for stars and the Sun. Convection is the transfer of heat by the bulk transport of heated matter, whereas radiation is the direct transfer of energy at the velocity of light by electro-magnetic waves, such as radio waves, light or X-rays.

The photons liberated in the central nuclear powerhouse are high-energy gamma rays. These bump their way along, making countless collisions with electrons and nuclei. This bumbling process gradually increases the number of photons and lowers their average energy as they diffuse outwards from the core: first to X-rays and extreme ultraviolet, then ultraviolet, and finally to visible light. Out to about 1 million km (almost nine-tenths of the distance from the core to the photosphere), radiation is the principal means of energy transport in the Sun. Convection only takes over in the final steps to the photosphere. The fact that convection does not occur within or close to the core has an important consequence: the ashes of nuclear burning do not get stirred up and mixed with the Sun's topmost layers. Therefore the solar atmosphere, which we can see directly, is not contaminated by nuclear waste from the Sun's own reactions but still has the same composition as the youthful Sun of 5 billion years ago. This provides us with an important source of information on the chemical composition of the early solar system.

As the photons approach about the last one-tenth of their journey an important change sets in. The pressure, temperature and density all decrease from the core to the photosphere. Around 1 million km out, the gas properties have changed so much that convection sets in. What has happened is that on the imaginary journey from the core we have encountered at last atomic nuclei that are cool enough, and therefore travelling slowly enough, to be able to hang on to electrons. Partially complete atoms, called ions, are therefore formed. These ions have a drastic effect on the progress of the photons, which, by the way, are now ultraviolet rays rather than ghastly gamma rays.

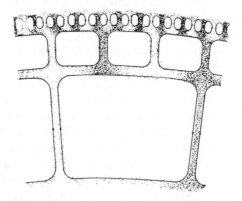

The hierarchy of convection cells in the Sun. As with most diagrams of the layered Sun this is schematic rather than exact.

The ultraviolet photon can be readily absorbed by the ions in the Sun (the high-energy gamma rays if they still existed would simply smash them to bits), so quite suddenly the bulk solar material is in the right condition to absorb the radiation, as opposed to kicking it around from nucleus to nucleus. In other words, the material at this distance has become completely opaque. It seethes with trapped energy.

The gas responds to the energy dumping by bubbling furiously. The whole situation is now locally unstable, leading to turbulent convection as the opaque gas which is blocking the radiation is forced to rise up through the Sun to cooler layers. Convection is a highly efficient way of moving energy around inside stars, so our photons make the last part of the journey to the photosphere very swiftly. The convective zone extends from a depth of around 150,000 km (100,000 miles) up to the photosphere. As mentioned previously, the photosphere is the transition zone where the Sun becomes very transparent to visible light, and this is the form in which much of the energy chooses to leave the Sun.

Theoretical astronomers think that the convection zone is possibly arranged as a series of three layers of convective cells. Deepest of all are the giant cells, each 150,000 km in diameter. Then an intermediate layer of cells carries the energy up to the frothy photosphere. Here a layer of small currents a few hundred kilometres across and something over 1,500 km deep reaches the surface. The bubbly top of this upper layer is the Sun's visible surface—at last the photons can zip away unimpeded (well, *almost* unimpeded) on a cosmic journey that will last countless aeons for most of them. Less than one photon in a billion will manage to hit the Earth.

Under excellent observing conditions, professional telescopes can photograph the fine mottling of the solar surface, caused by the convection cells. This seething surface structure is called granulation. Some of the finest photographs of it have been taken by telescopes taken to very high altitudes in balloons. (Under no circumstances should you attempt to see the granulation by looking directly through a telescope.) The foaming pattern changes ceaselessly as individual granules form and dissolve in an interval of just a few minutes. Measurements of the speed of gas within a granule have shown that the bright centre of a granule is made of gas floating upwards whereas the dark boundary is cooler descending material.

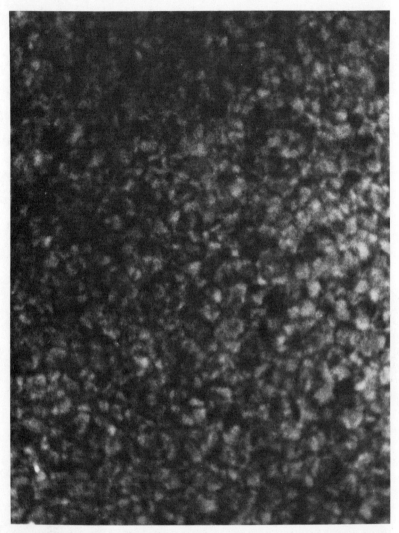

The mottled surface of the Sun with its cells of granulation. This is the bubbly froth of the photosphere, a layer in ceaseless motion. Each bubble is about as large as the British Isles. (Hale Observatories, California)

As the convection churns up the photosphere, important physical effects take place in the chromosphere. In a sense the chromosphere is a gassy froth sitting on the photosphere. All this bubbling of the photosphere fills the chromosphere with activity. The battering from below causes pressure waves—sound waves—to rush through the chromosphere. As they do so they heat the gas by speeding up its individual atoms. The sound waves generated by the boiling din in the photosphere are partially absorbed in the chromosphere. This absorption of mechanical energy in part accounts for the rapid rise in chromospheric temperature from about 4,500 K to 1 million K; the temperature rises at the rate of one degree every 2 metres within the chromosphere, the base of which is the coolest part of the Sun's outer layers. In addition to the sound waves, magnetic and gravity waves triggered in the photosphere also dump their energy in the chromosphere.

Once the million-degree gas at the top of the chromosphere is reached, the out-flowing energy has encountered the corona. Within

The spectrum of the chromosphere flashes into view during total eclipses. The bright arc-shaped lines are emission lines in the chromosphere. They match absorption lines in the Fraunhofer spectrum of the photosphere. It is only during eclipses that we can readily perceive the light being re-radiated by the atoms responsible for absorption. The American astronomer Charles A. Young discovered the reversed spectrum of the chromosphere during the 1870 eclipse visible from Spain. (Hale Observatories, California)

the corona the temperature is in the range 1–2 million degrees, making this the hottest part of the Sun that is readily observable. It is a strong source of X-rays but a very feeble emitter of visible radiation. This hot corona is not firmly anchored into the Sun's gravitational field. Its outer fringe is streaming into space—the solar wind.

The breeze of particles, or solar wind, has been studied by space probes. It almost certainly receives its outward push from the wave energy being pumped up from the Sun's convective regions. Now the particles can stream out through interplanetary space, taking a small proportion of the solar energy flux with them.

Our examination of solar energy flow from the nuclear furnace to the frozen universe has required a substantial knowledge of the architecture of the Sun: its various layers, their densities and temperatures, and an overall picture of the sources of stability. It is worth emphasizing again that the Sun is the only star from which it is possible to measure some of the quantities involved. Astronomers cannot,

The upper chromosphere is captured in this picture made in the light of ionized calcium. The white area shows a region of extra solar activity. The fine bristles are termed spicules. They mark out a network of supergranular cells each of which is 30,000 km across. (Bruce Gillespie, Kitt Peak National Observatory, Arizona)

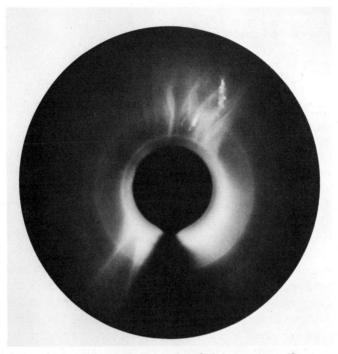

The solar corona photographed by the Skylab coronagraph. A spray of material from an active region is sweeping up through the corona to join the general outward flow that is the solar wind. (High Altitude Observatory, Colorado)

for example, detect particles streaming out in the stellar winds of nearby suns. Stellar coronas have now been detected by X-ray telescopes. The source of solar energy, nuclear reactions, had to be discovered before the energy sources of stars in general, and peculiar stars for that matter, could be investigated. So it is legitimate to ask: how certain can astronomers be that the picture given in this chapter is correct? The fair answer is that they are certain to a high level of acceptability, but any scientist would be foolish to claim complete certainty on so complex an object as our Sun. As we shall see in the next chapter there are still lingering doubts, particularly about the validity of our models of the solar interior.

The Puzzle of
Solar Neutrinos   6

Scientists generally work with models: simplified accounts of how nature works. An engineer will frequently study the behaviour of scale models that arc miniature versions of the real thing. For complex structures the engineer uses a computer to model the behaviour of structures and mechanisms. Astronomers cannot build scale models: their models of the universe, of galaxies and of stars are purely theoretical and will always remain so—the sheer size of the objects to be modelled ensures that! Scientific models have several purposes. For a model to be judged a success it must give an adequate explanation of the phenomena it seeks to elucidate. To find another analogy we can turn to the laws of universal gravitation. Newton's laws give a satisfactory result for the fall of an apple or the swing of a pendulum; the results are accurate for everyday purposes and therefore they represent an acceptable explanation of falling apples and swinging pendulums. Newton's laws do not, however, give a completely satisfying answer to the motion of planet Mercury in the Sun's gravitational field. For this we need a more complicated model—Einstein's General Theory of Relativity. This theory is able to explain why the elliptical orbit of Mercury is steadily slewing round, the effect generally referred to as the *precession of the perihelion* of Mercury's orbit. But even Einstein's theory does not give the complete picture because it leads to paradoxes or implausible results when one considers the behaviour of the ultimate gravitating objects, black holes.

Now, historically, the breakdown of a hitherto successful model has frequently led to much improved versions. Newton replaced Kepler, Einstein replaced Newton, and so on. Therefore the failure of a previously successful model may occasionally set scientific inquiry on completely new and fruitful paths.

Until very recently it seemed that one great area of astrophysical theory, the structure and evolution of the Sun and stars, had been completely dealt with. The mathematical and computer models appeared to give a watertight explanation of how real stars worked. But one issue in particular niggled the theorists: since we cannot directly observe the inside of any star how can we be so sure that the models are indeed a good approximation to the reality of the natural world? In the case of our Sun, it is important to remember that measurement of its energy emission today is telling us about conditions in the nuclear reactor millions of years ago, when the photons we now collect with our bodies and telescopes were first created. A prediction of the existence of a nuclear particle, made in 1931, may, however, be applied to teach us about conditions in the Sun today.

In the pioneering days of nuclear physics Wolfgang Pauli was puzzling over the radioactive decay of nuclear particles. He concluded that something more than the mere emission of electrons must be going on in these decays, for he could not make the outcome of the decay add up properly. To satisfy conservation laws—statements that certain physical quantities can neither be created nor destroyed in reactions—he had to postulate the existence of a new sub-atomic particle. This undetected particle he called the neutrino—or 'little neutral one'—which we briefly encountered in Chapter 4. This particle of the sub-atomic world is most remarkable for it has no electrical charge, its mass is negligible and it travels with a velocity equalling light. Neutrinos are just bundles of energy, but they are utterly different in character to the photons, which can be viewed as packages of electromagnetic energy. Neutrinos are amazingly penetrating because there is almost no way they can interact with everyday matter once they have been created. Countless millions of them are passing through you every second. Even a slab of lead stretching from here to Pluto would have scarcely any effect on the intensity of a neutrino beam. The very low chance of any interaction with matter meant that a quarter century elapsed between Pauli's prediction of the neutrino and definite detection of one in an experiment.

The Sun is full of neutrinos: the reactions inside the Sun release two neutrinos for every fusion of a new helium nucleus (Chapter 5). During every second the Sun's nuclear fires release 2×10^{38} neutrinos. What happens to them? Since any selected neutrino is most unlikely to interact with matter they all stream straight out because matter, even inside the Sun, cannot hinder their progress. So now we have an interesting situation: today's electromagnetic energy will not get out until millions of years hence, whereas the neutrinos zip away

almost unimpeded at the speed of light. Even by the time they encounter the Earth, which is a small target at a large distance, a
million million pass through an area the size of a large postage stamp
(say 10 square cm). We are all utterly unaware of this vast flux of
harmless particles flowing through us all the time.

But if a detector of solar neutrinos could be made, scientists could
look at conditions within today's Sun. They would be able to compare
the predictions of the solar model-makers with a real contemporary
measurement. Raymond Davis, an American physical chemist, has
devised just such a neutrino detector.

Davis did not set out to make a Sun-penetrating telescope. He
worked on an altogether different problem, apparently unrelated to
astronomy. His starting point was the prediction that there are really
two classes of neutrino: neutrinos and *anti*neutrinos, the latter being
the mathematical opposite of the former. Although our everyday
world is composed entirely of matter, the sub-atomic world is a mix-

*The neutrino telescope of the Brookhaven National Laboratory is operated
by Raymond Davis. The main detector is a tank holding almost half a
million litres of dry cleaning fluid. To cut down the cosmic ray flux the
tank is deep in a gold mine in South Dakota. The cavity housing the tank
was specially excavated by the mining company. (Brookhaven National
Laboratory, New York)*

ture of matter and anti-matter. When a particle collides with its precise anti-particle they both vanish in a puff of energy ($E = mc^2$ once again!). Davis wanted to distinguish neutrinos and antineutrinos and to do this he took up a suggestion of Bruno Pontecorvo, who was working (1947) at the Chalk River Nuclear Laboratories in Canada. Pontecorvo had proposed the use of atoms of chlorine-37 to make a neutrino-catcher. The reasoning behind this was that *if* an atom of chlorine-37 could grab a neutrino of the right energy it would be transformed to an atom of argon-37, an electron being kicked out in the process. Antineutrinos can never have this particular effect on chlorine-37; therefore, the argument ran, to detect neutrinos start with chlorine-37 and see if any of it gets turned into argon-37. Davis first attempted this task right next to a nuclear reactor on Earth, where a large flux of neutrinos was expected.

To carry out the experiment it's necessary to measure the amount of argon-37 produced from chlorine-37. This particular form of argon reverts to chlorine-37 in an average time of thirty-five days, emitting an electron, which carries a precise amount of energy as it does so. This decay provides the means of detecting the argon, through the recognition of the energy fingerprint in the electrons kicked out. For this detective work at least ten argon atoms are needed.

In his pioneering 1955 experiment Davis used 15,000 litres (3,000 gallons) of carbon tetrachloride (perchloroethylene, a cleaning fluid) because this fluid is rich in chlorine-37. After some days argon-37 was present. He had detected neutrinos from the reactor and also placed a crude upper limit on the flux of neutrinos from the Sun.

Successive improvements in the basic scheme eventually led to the construction of a solar neutrino detector tens of thousands of times more sensitive than the original equipment. The solar neutrino telescope is, essentially, a tank of cleaning fluid—it holds 450,000 litres (100,000 gallons), as much as a 25-metre swimming pool—and a means of picking up the tiny amount of argon-37. The detector must be shielded from cosmic rays which would form argon-37 in various ways, so it is located a mile underground, at the bottom of an old mine in South Dakota. The concept of an astronomical telescope deep underground has intrigued professional astronomers and the general public.

The Davis machine cannot detect every kind of neutrino from the Sun. Only the neutrinos made in the decay of boron-8 (^8B) have

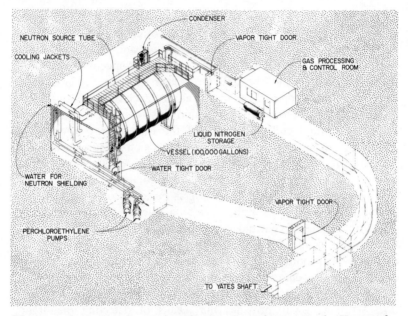

The general arrangement of the solar neutrino telescope in the Homestake gold mine. The apparatus is nearly 1,500 metres (4,850 feet) below ground. (Brookhaven National Laboratory, New York)

precisely the right energy to kick the chlorine-37 across to argon-37; the other solar neutrinos are really ineffective. As mentioned in Chapter 5, the boron decay features in the third of the three possible branches of the proton-proton cycle. Now we come to the crucial point: the rate at which neutrinos are made through this route, as opposed to the other two, depends very strongly on the temperature (approximately as T^{13}). Hence, measurement of the neutrino flux gives us a way of taking the temperature of the Sun's deepest layers. In turn, the discovery of this temperature provides an important and independent check on the techniques used to make theoretical models of the workings of our Sun.

Over the years Davis and the theorists played a remarkable game of intellectual leapfrog. Each time he improved the sensitivity of his telescope, the theorists would give a downward revision of the expected neutrino flux. The number of neutrinos detected is far less than anticipated, and yet the measurement can scarcely be in error by more than 10 per cent. The generally accepted ideas of how the Sun works give a neutrino flux ten times higher than the observed flux. So, the theorists have tinkered with the solar models to see if a little fine-tuning would bring prediction into line with experiment.

One attempt to explain away the discrepancy has been made by Dilhan Ezer and Alastair Cameron. They have suggested that if the Sun's outer layers have in the past somehow been mixed with the inner core, then the composition of the nuclear reactor will have been altered. This mixing can bring fresh supplies of helium-3, which immediately flares up the reactor. The extra energy production then causes the core to expand, and the temperature to fall; and as the temperature drops the neutrino flux decreases catastrophically. Periodic stirring of the Sun in this way could occur every 100 million years or so, and it would lead to a lower core temperature and diminished neutrino flux for around 10 million years.

The Ezer–Cameron approach is just one example of how judicious fiddling with the accepted model can alter the neutrino flux. A related idea is to have a rapidly spinning core, which would perhaps trigger the mixing. A more novel suggestion is that the Sun has a small black hole at its heart, but this speculation has not been taken seriously by solar physicists.

Whichever way one looks at the solar neutrino experiment there is a problem. Davis detects hardly any neutrinos from the Sun, whereas all believable models of the solar interior predict at least a few. Furthermore, standard ideas on how stars work certainly lead astronomers to expect a neutrino flux that should be easily detectable by telescope. This is the crucial point: does it imply that our Sun is not a completely normal star? Or is it going through a very long-term mixing process?

The result of the neutrino experiment has important consequences for our understanding of the behaviour of the Earth's climate. The absence of neutrinos may mean that the central temperature of the Sun is currently cooler than average, and the solar luminosity lower than average. Possibly this lowering of the Sun's temperature is a trigger for the major ice epochs which occur at intervals of 200–300 million years or so. If this reasoning is correct then the Davis experiment is showing that we are currently experiencing an ice epoch; the Earth is relatively warm right now only because we are in a short (200,000-year) interglacial period.

Lifespan of Our Sun

7

If the Sun were the only star in the sky we would not be able to deduce much about its birth, life and death. Only by observing many other stars have astronomers been able to give an outline of their life histories in general and of the Sun in particular. The analogy of growth in a forest is useful here: no botanist could sit down and watch a tree germinate from a seed, grow into a sapling, mature into a fine tree, and then die, because most trees live longer than most botanists! However, the observation of trees of various ages and populations is enough to enable the botanist to work out the life cycle of trees. So it is with the stars: astrophysicists have learned in the last half century how to recognize young ones, the vast numbers of middle-aged ones like the Sun, and those in the final stages of their evolution. This chapter gives a brief sketch of the past and future history of our Sun.

First it is a good idea to establish the present ages of the Sun, our Galaxy, and the Universe, in order to fix the timescale of the history. The age of the solar system is found by measuring the age of the oldest materials we can lay our hands on. There is no primeval material on the surface of the Earth. Continental drift, the weather, oceans, and ice ages have moulded the surface rocks to such an extent that they no longer retain any information on the age of our planet. More seriously, though, the Earth did not separate into a solid rock crust, a mantle, and a liquid core until many hundreds of millions of years after it formed. So, information on the origins of the solar system must be sought elsewhere: in meteorites and on the Moon.

Planetary scientists state that meteorites are fragments of rock left over from the early history of the solar system. They probably formed not long after the Sun had assumed its distinct identity. When this happened some of them managed to trap minute traces of radioactive elements. All through subsequent history these radioactive substances have decayed—some types quickly and others extremely slowly—with the result that the radioactive elements gradually decline and their decay products build up. A meteorite is a cosmic clock that is gradually running down. By selecting a radioactive element and carefully measuring the relative amounts of the isotopes that have yet to decay—relative to the decay products of those that have already decayed—scientists can date a meteorite. There are several complications: for example lead is the decay product of uranium and thorium, but lead isotopes will have been present as well as uranium and thonium, so the ashes of uranium-thorium decay are already contaminated. Apart from the uranium-thorium decay, radioactive clocks from the decays of potassium to argon, rubidium to strontium and, recently, samarium to neodymium, can be read by meteorite scientists. This burgeoning science of *cosmochronology* has come up with a remarkable result for meteorite ages. They were all made about 4.57 billion years ago, within an interval of around 30–100 million years. This, of course, is strong evidence that the Sun and solar system date from around 4.6 billion years ago.

Radioactive clocks also make it possible to determine the age of Moon rocks. One of the main reasons for wanting to have samples of lunar material on Earth was to see what time the lunar clocks are showing. The age of the Moon inferred in this way is 4.5 to 4.6 billion years. Incidentally, the most ancient rock samples on Earth are around 3.6 billion years old (in West Greenland), and their lead composition is consistent with an Earth age of about 4.45 billion years.

The conclusion is that the solar system formed between 4.5 and 4.6 billion years ago. Very much greater uncertainty attaches to the ages of our Galaxy and the Universe as we know it. At present the Galaxy is believed to be about 10–12 billion years old, whereas the origin of the Universe dates from 13 billion years or earlier. At any rate we can be certain that the Galaxy did not condense until, at the very least, a few hundred million years had elapsed in the young Universe; and that the Galaxy was already at least 5 billion years old before the Sun condensed.

Ideas on how stars, and hence the Sun, come into being have advanced greatly in recent years. Clouds of gas in interstellar space are the birthplaces of new stars. The formation of a new star is a long and slow process when measured on a human timescale; almost all the stars visible to the naked eye were already there in the sky before mankind walked the Earth. And yet we know that new stars must grow inside gas clouds. For a start, young massive stars, which only live for a few tens of millions of years, are usually formed in association with clouds of hydrogen, helium, and other elements. Also, new stars have turned on in the Orion nebula, one of the nearest stellar nurseries. Finally, the chemical make-up of the interstellar gas is very similar to that of the Sun and stars: roughly three-quarters hydrogen to one-quarter helium with a liberal sprinkling (2 per cent) of the heavier elements.

The life style of a gas cloud depends on balancing gravitational forces against the pressure resulting from heating and compression. We have already encountered one version of this eternal cosmic battle in our discussion of the Sun's stability. The gravitational forces want to squeeze parts of the cloud together, whereas the thermal energy is trying to waft it into space. Half a century ago the Cambridge theorist Sir James Jeans worked out the conditions under which a clump of gas could collapse down until it became something useful. The conditions for compression—and hence the birth of stars—depend on the temperature and mass of the gas: cooler clumps of gas do not need to be as massive as warmer clumps in order to shrink. Nevertheless, even at a mere ten degrees above absolute zero a rather dense cloud of gas and dust needs to weigh in at tens of solar masses if it is to contract. Warmer clouds need to be more massive still. Jeans's theory accounts for the fact that stars are generally born in families, called clusters. A typical brand new cluster will have thousands of sun-masses of stellar material, forming perhaps a couple of hundred stars. It's much easier for the interstellar gas to make 200 stars than to make, say, two dozen, because the gravitational crunch is more effective in the bigger clouds.

As you can see, it's not easy to make stars from interstellar gas. If it were trivial then presumably all that gas up there would have turned into stars long ago—before the Sun was made! The cloud that spawned our Sun had several problems to solve: basically it started out too hot, spinning too fast, and containing too much magnetism to make stars. As the cloud collapsed it warmed up. You may have noticed that a cycle pump heats the air being forced into a tyre. The collapsing interstellar cloud has to get rid of the heat, or else the

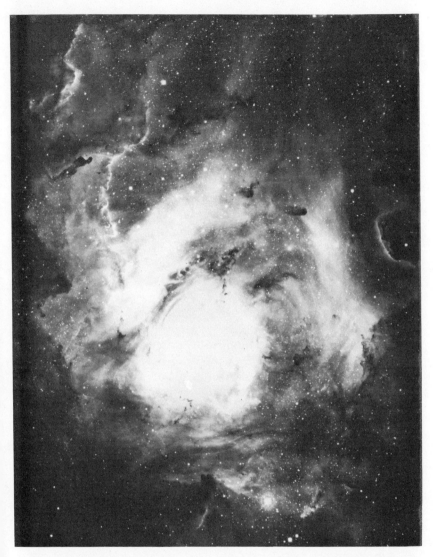

The Lagoon Nebula (Messier 8) is a large collection of cosmic gas and dust. The dark clumps of matter at the edge of the nebula are typical of the regions in which star formation occurs. Gravitational attraction squeezes these condensed cloudlets together. They break into smaller lumps as they shrink. Some of them collapse down until they are dense enough to ignite nuclear burning in their centres. Thus is a new family of stars born from the interstellar medium. (Kitt Peak National Observatory, Arizona)

shrinkage will just cease. One thing that probably helps the collapse is the Galaxy itself, interestingly enough.

Our Galaxy has two spiral arms, and out where the Sun is it takes about 250 million years for gas and stars to make one circuit. Periodically (let's say every 125 million years) a given cloud passes through a rotating arm. As it does so it gets a shock, and considerable compression takes place as it crashes through the denser part of the arm. This alone will trigger the contraction. Nearby galaxies, studied by optical and radio astronomy, clearly demonstrate that star formation is most vigorous along the edges of spiral arms, where material is piling up. So, about 5 billion years ago, the sun-cloud got a massive wallop as it tried to tunnel through a spiral arm. As the cloud collapsed it heated dust within itself, and this got rid of gravitational potential energy as infrared radiation. Sites of star formation are very dusty places and the heat radiated by them features prominently in the infrared sky. The sun-cloud may also have rid itself of excessive energy through the radio emission of molecules. Radio astronomers have found that some molecules, for example water, can pump microwave energy away from dense clouds extremely effectively. Essentially the water vapour functions as a maser (similar to a laser, but emitting microwaves) and this drains off energy as the collapse proceeds. In summary, the collapsing cloud has to pump out energy in the infrared or microwave regions of the electromagnetic spectrum.

I said above that the Jeans criterion implies that the cloud which seeded the Sun may well have been thousands of solar masses. As it condenses it fragments in cloudlets because parts of it become unstable. These cloudlets in turn will shatter as the collapse continues. Finally there remain dark protostars and the proto-Sun, still collapsing but now close to their final identity. To reach this stage took about 400,000 years in the case of our Sun.

What happened next is not entirely clear. The proto-Sun was not hot enough at this stage to sustain nuclear reactions. It also had to live with the rotation problem: having shrunk down it would have been spinning quite fast, just as a child sitting with extended arms on a spinning stool will speed up if he then pulls his arms in. Possibly the rotation slowed somewhat as the magnetism within the proto-Sun got enmeshed with the galactic magnetic field. Magnetic-field lines are a bit like elastic; you can stretch them out but they resist more and more as you do so. In this way the rotating proto-Sun could have wound up the local magnetic field, which in turn would have acted as a brake on the rotation. Furthermore, the combination of rotation and magnetism would have helped to form a disc as the cloud approached the size of the Sun.

The final stage in the Sun's birth took about 100,000 years. Inside the proto-Sun cloud, a core formed and this collapsed under the force of gravity, increasing in temperature and pressure all the time. The first nuclear reaction started up, combining lithium-7 with a proton to form two nuclei of helium. This reaction occurs at temperatures of around 1 million degrees, and is soon used up. The lithium is a kindling for the nuclear furnace. With the onset of nuclear reactions the isolated fragment of gas had finally become the youthful Sun, but it still spent some time settling down. In the first few million years the nuclear reactions built up steadily. This phase was complete after about 50 million years for the Sun. Meanwhile, it probably cast off a fierce wind, much stronger than today's solar wind, as it expelled the diaphanous fringes of its parent cloud.

While the nuclear furnace was being organized, the debris left behind during the final collapse, to a solar-sized ball, came together in clumps to form meteorites and proto-planets. By reading the meteorite clocks we know that the condensation, or freezing, of these rocks took place over an interval of about 30–100 million years. The flattened planetary disc was linked by magnetism to the Sun, and this connection may have helped to slow the Sun's rotation a little. However, the main way in which the Sun lost its spin (it now takes a month to rotate once) was by flinging away angular momentum or rotation energy with the great gale of a particle wind.

The coalescence of the planets must have taken a few hundred million years. In the last stages small rocks rained down violently on the bigger rocks; this can be seen on the surfaces of Mercury and the Moon particularly where the ancient landscapes still bear all the scars of a bombardment from a fearful cosmic battery. Lumps of rock—meteoroids—still wander, if speeds of 30–100 km per second qualify as wandering, through interplanetary space. When you see a shooting star you're seeing the one-second death of a rock that is older than anything on the surface of the Earth—a rock left stranded in the solar system 5 billion years ago. Sometimes fragments hit the Earth; these are the meteorites.

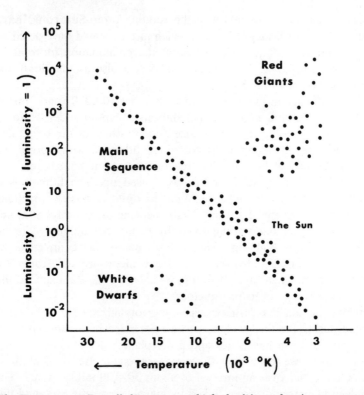

The Hertzsprung-Russell diagram in which the life cycle of a star may be followed. The graph may be presented in more than one form and is here shown as a plot of surface temperature against luminosity. For most of its life the Sun will be stationary on the main sequence, and it will probably then progress to a red giant and finally a white dwarf.

I have already described the source of the Sun's energy throughout the greater part of its life. It is now truly middle-aged, in that about half the hydrogen fuel in the core is already used up. So there are another 5 billion years to go! If we represent the first half of our Sun's life by one of our calendar years, the emergence of mankind can be said to have occurred at around 22.00 on December 31, the start of civilization at ten minutes to midnight, and the invention of the telescope in the last second. Personally I doubt if any direct descendants of ours will be troubled by the Sun's finite, but huge, future lifetime. It seems extraordinarily vain to assume that we are the *immortal* peak of an evolutionary chain; we've just reached a higher pinnacle than anything else so far, as judged by our standards.

Anyway, what will life of any kind experience when the Sun has aged another 5 billion years? The schematic way of plotting the end is by means of a Hertzsprung-Russell diagram, which has the surface temperature of the Sun along the horizontal axis, and its brightness plotted vertically. The Hertsprung-Russell diagram is particularly useful for following the rapid changes that will take place when our Sun is battling against gravity in the face of ever-dwindling energy reserves.

At present the Sun is located on the main sequence, that is to say, the band in the diagram where all hydrogen-burning normal stars are found. Stars to the left and above the Sun's main sequence location are more massive than the Sun, those below less massive. It's vital to understand that neither the Sun nor any other star moves up or down the sequence. It stays in almost the same place as it relentlessly consumes hydrogen, moving only a little to the right and above the main sequence, as the helium replaces hydrogen. The observed effect of this is that as the Sun feels its age, but while it still has a supply of hydrogen, its luminosity will increase by about one-quarter, although the surface temperature falls a little. The Sun, in short, becomes a little larger, somewhat redder, and feels considerably warmer. This phase of the evolution will, of course, be profoundly important for any life that is on the Earth's surface a few billion years hence. The reason why the Sun hots up is interesting: the burning of hydrogen to make helium consumes electrons—two of them for every helium nucleus made. So, as time goes by, the electrons become depleted, and this makes it easier for the energy to get out of the Sun's core, simply because electrons take prime responsibility for holding up the photons; they are the prime cause of the Sun's opacity.

When the nuclear reactor is finally burnt out, the Sun's core will contract. Again, the physical cause is very simple: once the energy flow ceases the core cools, and to counteract this loss it shrinks, releasing instead the potential energy of its own gravitational embrace. Core shrinkage means that unburnt hydrogen outside the core falls closer to the centre of the Sun. This will then turn on a new energy source: hydrogen raining down on the ashen centre gets compressed and heated to such an extent that hydrogen-burning is ignited in a shell around the core. As shell burning proceeds, helium is dumped into the core on the inside of the shell, while fresh hydrogen is consumed on the outside.

Dramatic changes will be seen by (imaginary?) astronomers in other planetary systems as they view our Sun 5 billion years hence. As the core shrinks, the outer layers greatly expand, leading to a huge red Sun at least ten times the diameter of today's Sun. Such a star is in the red giant phase of the HR diagram, emitting energy at thousands of times the normal rate, and searing the planets. Subtle physical effects are at play here. The collapsing core stays at more or less the same temperature. To do this it has to get rid of some of its internal energy, which is instead transferred to the outer layers, or envelope, causing dramatic expansion. A paradoxical golden rule in stellar evolution is that when insides shrink, outsides expand. The other effect is that a cooler redder Sun will give the planets a very hot time. This is because the Sun's surface area, and hence its apparent size, increases at least one hundredfold and this over-compensates for the lower luminosity per unit area. What will happen on Earth? The oceans and rivers will boil, the ice caps melt, and the atmosphere will be whisked away into space. The barren, rocky face of our planet will, at the same time, be blasted by the fury of the solar wind, now flowing once again at a much faster rate. Every year the red giant Sun will lose one-millionth of its mass. In the outer solar system the icy giants Jupiter and Saturn will be released from their 10-billion-year deep freeze. Icy mantles of methane, ammonia and hydrogen, tens of thousands of kilometres thick, will evaporate as well, as the Sun finally warms the outposts of the solar system. Maybe these planets will be stripped right down to rocky cores; we don't know. And in the inner solar system, planet Mercury will be orbiting almost inside the Sun.

There are other effects: far across the Galaxy the Sun will become visible as never before. In our language it will have increased in apparent magnitude by six. Just as red-giant Betelgeuse is currently one of the brightest stars in our sky, so the Sun will dominate the alien skies of unknown planets deep in space, a thousand light years away.

In astronomical terms this glorious expansion will be short-lived. Like most sudden growth phenomena in nature, the Sun will over-react, will over-reach itself. The high luminosity will be an unsustainable drain on the now limited fuel reserves. The shells will burn hydrogen faster and faster, heaping helium ashes upon the core, and as the central slag heap grows in mass its temperature will rise. When it reaches about 100 million degrees the helium nuclei will start to fuse in triplets to make the nuclei of carbon atoms. The nuclear cooking will run away uncontrollably, for the temperature will increase further as the reaction proceeds and this will only push along helium-burning with even greater enthusiasm. The conditions will be quite unlike main sequence living in which extra heating is balanced out by a pressure and modest expansion—an important stabilizing influence. No, the endpoint of solar evolution is quite uncivilized because the pressure in the core is insensitive to temperature. This will be the peak of red giant evolution, a point at which the helium flash takes place, forming carbon (and nitrogen) in the core. The final burst will push the Sun's outer boundary possibly as far as the Earth's present orbit.

Contrary to popular belief, our Sun will not become a nova or supernova star. The helium flash, for example, does not produce a sudden brightening or give the appearance of a stellar explosion. This is because the view of the core, where it's all happening, is shielded by the extensive envelope.

The subsequent behaviour of our Sun cannot be predicted with great certainty. As it falls back in on itself, at the end of the red giant era, it may become unstable. If a helium-burning shell is established at the edge of the core, frequent flares will take place. The Sun would then shrink and become hotter, until the pressure of radiation flowing from the dying core might be enough to lift the atmospheric veils back into space with a velocity of a few tens of kilometres per second. We do not know whether the Sun will lose its envelope all in one go or in a series of bursts. But what is pretty certain is that stars with the mass of our Sun produce planetary nebulae, in which a ring of glowing hydrogen and helium surrounds a tiny very hot star with a surface temperature of 100,000 degrees. This central star is in fact the core, the hot but dead nuclear power-station of a star.

The final fate of suns is remarkably simple. The nebulous envelope gradually drifts into interstellar space, where presumably it is eventually incorporated into the gaseous nebulae from which another generation of stars and planets may be born in the far future. Meanwhile the hot core steadily cools down as it radiates into space. All the time it is moving downwards in the HR diagram and is to the left of the main sequence. It is now a white-dwarf star—a good fraction of a solar mass packed into a ball the size of the Earth. The white dwarf does not collapse even though its internal gravitational force is high. Electrons, those tiny charged particles that form the outer clouds of atoms, are pressed so tightly together that they can exert a pressure high enough to balance gravity. This electron pressure does not originate from the electric repulsion of the electrons as you might anticipate. Rather it is a special force that has no counterpart in the world that we humans experience directly. It is a quantum force, a force of the micro-world of elementary particles, which arises because the electrons inside the star must all have different energies.

As the white-dwarf Sun continues to cool it will slowly fizzle out. Eventually it will be too cold to emit any radiation as visible light. It will become a black dwarf, an almost undetectable heap of nuclear waste, composed essentially of helium, carbon, nitrogen and oxygen. All the matter it contains has at this stage reached the end of the cosmic road as far as the predictable future of the Universe is concerned. Six billion years from now the Sun and Earth will definitely be dead.

A study of the evolution of stars heavier than the Sun takes us back to the conception of the solar system by a roundabout route, as we shall see. It is not appropriate, in a book about the Sun, to go into the finer details of stellar evolution, so only the main points are outlined. Stars more massive and therefore hotter than the Sun process their hydrogen to helium in a different manner, through the carbon-nitrogen cycle. There are six stages in this reaction, which uses carbon and nitrogen nuclei at intermediate points without actually consuming them, to make helium from the lightest element. Inside the Sun the temperature is not high enough for the carbon-nitrogen cycle to work very impressively, but it does operate inside hotter stars, where the central temperature exceeds 16 million degrees. The carbon-nitrogen cycle is only contributing 2 per cent or so of the Sun's energy.

One of the most important effects in stars more massive than the Sun is their shorter lifetimes. Now this may seem a little odd: surely the bigger stars have more fuel and ought to last longer? Well, they do have more resources, but they get through them much faster. To give an example: a star five times the mass of the Sun has about five times as much hydrogen to put through the reactor. But the higher mass

SUPERNOVA IN IC 4182

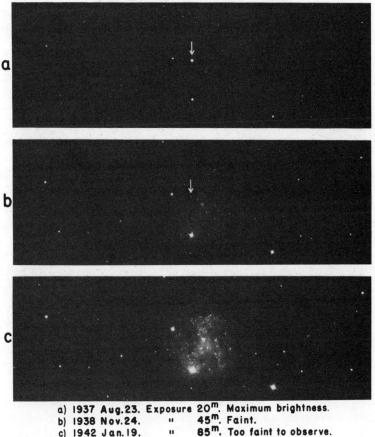

a) 1937 Aug.23. Exposure 20^m. Maximum brightness.
b) 1938 Nov.24.   "   45^m. Faint.
c) 1942 Jan.19.   "   85^m. Too faint to observe.

A supernova explosion in another galaxy, the spiral IC 4182. In the first exposure on 23 August 1937 the star that has exploded shows up clearly but no other part of the galaxy has been recorded by the photograph. The second exposure was made three months later; the dying supernova is barely visible. Finally a long exposure five months after maximum brightness reveals no trace of the star. (Hale Observatories, California) causes a tighter central squeeze, and hence higher temperature and pressure at the centre; therefore the rate of nuclear burning shoots up by about one thousand times. Five times the fuel eaten at one thousand times the Sun's rate means death two hundred times earlier—after about 50 million years or so. Now ask yourself about the origin of life: it took 5 billion years to go from the proto-Sun to Man; could intelligent life evolve on planets round the massive stars? Probably not, because their life is too short.

Yet, in a truly amazing way, these prodigal stars have made life as we know it possible. When the Sun dies it will throw a shroud of gas into space and shrink to a ball of lightweight elements: helium, carbon, nitrogen. But when a massive star wants to die it gives in far more spectacularly. An object ten times as massive as the Sun simply cannot snuff out gently. When the fuel runs out the core of the star collapses very fast—perhaps in under a second—turning either into a tiny ball of neutrons 10 kilometres across or, maybe, even into a black hole. The effect of this implosion, as it is called, is to send a shock wave running through the outer envelope, as it starts to crash down on the core. This compresses the gas, and within a further second it is hot enough for a nuclear explosion. The crucial point is that the very sudden demise of the core, when gravity finally beats internal pressure, stops any civilized restructuring of the star. A dying massive star is bound to have a dramatic end in a supernova explosion.

The explosions of massive stars provide a breeding ground for heavy elements. The hot helium-rich environment cooks the nuclei to create isotopes of the elements such as neon, magnesium, silion, sulphur and so on, right up to iron. (These processes also take place in the hearts of massive stars.) Elements beyond iron are cooked up by capturing helium nuclei and, more importantly, free neutrons.

The supernova explosion casts gas into space, and this gas is greatly enriched by the heavy elements—some baked in the nuclear furnace and some toasted in the exploding atmosphere. This material shoots out at speeds exceeding 10 million kilometres an hour. Over millions of years it slows down and merges imperceptibly with the gas nebulae of the interstellar medium. Now the scene is set for a new stage of star formation, a stage in which a generation is enriched by the ashes of its forefathers. Thus, all the material on the Earth and in our bodies was made from hydrogen and helium in nuclear explosions

A negative photograph of an ancient supernova remnant, the Cygnus Loop. The wreckage is now probably 100,000 years old and it is gradually merging with the interstellar medium. (From the Palomar Sky Survey)

that took place long before the birth of the Sun. The implication is that the first generation of stars, made in the earliest history of our Galaxy, could not have had planets. Only stars made with a proportion of recycled material, stars like the Sun, can be accompanied by planets.

Our daytime star provides theoretical astronomers with a readily accessible testbed for their schemes, and indeed the Sun lies at the very foundation of models of star structure. Naturally if someone queries the status of Sun-models, waves of doubt and dread shudder through the ranks of the model-builders.

Just such a question led, ultimately, to the discovery that the Sun itself has the shivers. The remarkable way in which the shudders of the Sun were tracked down by Arizona scientist Henry Hill is yet another of those implausible-sounding stories in the history of astronomy.

The discovery of solar vibrations all started far away from solar studies with a new theory about the Universe, proposed as a rival to Einstein's general theory of relativity. This is the *scalar-tensor* theory of gravity, proposed by Brans and Dicke. It turned out, on analysing this theory, that the Sun could help in seeing if Einstein's theory really was inadequate. As mentioned already, the planet Mercury is an important probe of the Sun's gravitational field. The elliptical orbit of this planet slews round steadily in space, so that the orbit is a rotating ellipse that does not quite close back on itself as the planet completes a circuit. The slewing speed, termed the precession of the perihelion, has an excess component which is neatly explained by Einstein's general theory. Indeed, one of the triumphs of the theory was the very fact that it resolved all the difficulties associated with the behaviour of the Sun's closest planet.

In the newer Brans–Dicke theory of the interaction of matter and gravitation, 7 per cent of the excess movement of the orbit was left over. Accounting for that 7 per cent, in an attempt to save the theory, was to lead to new solar discoveries.

Bob Dicke thought of a way out of the dilemma: he said, suppose the Sun were slightly squashed, like an orange, then Mercury would not be travelling through a perfectly symmetrical solar gravitational field. Only a small amount of distortion, amounting to a difference in solar radius of a tiny 30 kilometres, when comparing the polar and equatorial radii, would be needed to rescue the new theory of cosmology by getting Mercury back on target.

Now measuring a distortion of 0.05 per cent in the Sun's shape is extraordinarily hard, and for theorists there was the equally knotty problem of seeing how the Sun might become mis-shapen in the first place. The excess oblateness, as the distortion is called, could be caused by rapid rotation in the core. This idea would imply that the core was spinning much faster than the Sun's outer layers. Such a notion was attractive for another reason: rapid rotation would lower the Sun's central temperature and, hey presto, decrease the neutrino flux. Things began to look grim for Einstein's theory as it seemed that the new theory could, incidentally, solve the neutrino problem. Only measurements could settle the issue.

The Sun does not have a completely blank face. Storms, flares and sunspots disfigure it; worse still they confuse shape measurements by messing up the brightness of the disc. At the University of Arizona,

Henry Hill built a telescope specially designed to detect distortion in the periphery of the Sun's shape. When the telescope set to work no distortion was found. The implication was that Einstein was right, the Sun is not spinning rapidly deep inside, and the neutrino problem is still unsolved. In an impressive series of measurements Hill and his colleagues went on to discover new effects: periodic oscillations of the Sun. At the solar limb he finds movement like the ringing of a bell. We can picture this as a very deep bass bell with a fundamental period of 52 minutes, together with several of the harmonic vibrations.

The discovery of the shivering Sun, a complete accident occasioned by an incorrect theory, has important consequences for the solar-model makers. Just as the seismic vibrations of the Earth, triggered by earthquakes, give information on the Earth's interior structure, so the Sun's normal modes of vibration are sensitively tuned to the run of temperature and pressure of the solar interior. Researchers in Cambridge, England, have stood Hill's marvelous data on its head and worked backwards to sort out the solar structure. In fact the classical models of the Sun withstand this test well, and we still have a neutrino problem.

Other groups of researchers have reported solar shivers of even longer periods of almost three hours. Since these measurements have not been confirmed by independent experiments they are not yet regarded seriously. But if a three-hour vibration really is bouncing around in the Sun the theorists would once more be thrown into turmoil.

The basic position today is that we think our knowledge of the Sun's interior is firmly based, with theory and observation reasonably interwoven. This also means that the ideas about the main sequence evolution of stars like the Sun are probably near to the mark. Of course, research on stars *per se* acts to reinforce solar studies. An important facet is the behaviour of a star cluster. When a family of stars is born in a cosmic gas cloud the members differ mainly in respect of their masses. Since the more massive stars evolve the fastest, at any later time the cluster gives a sketch of stellar evolution: the massive stars may be near death but the small stars have scarcely dented their fuel reserves. The range of star properties within a cluster can therefore be understood as a set of markers at different points in the evolution of a normal star. The clusters are the most important test ground for star modelling in general, and examining them has strengthened our faith in models for the Sun.

In this brief survey we have reached the frontiers of knowledge about the interior. Will new data resolve the neutrino question? Let's hope so. In the rest of this book our attention will be fixed on what might be called the solar exterior, the part that is amenable to observation.

Surface and Atmosphere

The sun's visible surface, the photosphere, is in ceaseless activity. Underneath the surface the turbulent rumblings of the convection cells boil endlessly away, creating the foamy structure of solar granulation, described in Chapter 4. The temperature of the surface can be defined in more than one way. For example, if the spectrum of the white light from the photosphere is matched as nearly as possible to the temperature curves of black bodies we can find the *black-body temperature*, the one that most closely mimics the spectrum of the Sun. This occurs for a value of 6,000 degrees absolute. Another way of finding a temperature is to ask how hot would a sphere the size of the Sun have to be, for it to radiate as much energy as we receive? Answer: about 5,800 degrees absolute.

There is no one 'correct' temperature for the Sun, because it is a complex object in which the temperature varies with height above the surface. We get energy from a layer which is about 500 kilometres, thick, the temperature of which varies with depth. The radiation received from the centre of the disc is mainly emitted by gas with a temperature of 6,500 degrees, whereas from the limb of the Sun it is rather cooler. Any method of defining an average temperature has to be a compromise, but this does not matter so long as the basis of assessment is given.

The Sun's sharp edge, mentioned on p. 59, arises in the following way. Near the photosphere most of the absorption of visible light is caused by a special kind of hydrogen atom. Normal hydrogen has one proton and one orbiting electron, an electrically neutral and stable arrangement. Sometimes the hydrogen can pick up an extra electron, on a temporary basis, to make hydrogen with two electrons and a negative electrical charge; this is called the negative hydrogen ion. Such a shaky marriage of atom and electron can only survive if the temperature is right. In the Sun the switch-over takes place very quickly, with the result that as the radiation percolates out it suddenly encounters the region where menacing hydrogen ions no longer survive, so we can see the radiation for the first time. As mentioned before, the suddenness of the change gives the Sun a sharp edge.

The yellow-white light from the photosphere has a smoothly varying spectrum without any lines embedded in it; but the light has to traverse cooler layers of atmosphere before leaving the Sun entirely. Within this cooler zone absorption takes place, absorption that tells us much valuable information on the atmospheric conditions. The layer is occasionally called the *reversing layer* in older textbooks.

That the rain makes rainbows from sunlight is well known. The scientific analysis of solar rainbows begins with Isaac Newton, who in 1665 spread the light into a coloured spectrum by using a prism placed in a narrow beam of light. He was experimenting in optics rather than making a deliberate astronomical observation; he found the coloured radiation and started solar spectroscopy. Wollaston was the first to record dark lines crossing the rainbow spectrum, in 1802. This finding spurred other astronomers to investigate the spectrum, the most prominent being Joseph Fraunhofer.

Fraunhofer, as part of a detailed study, designated the principal dark lines from red to blue-violet by letters of the alphabet and so, for the first time, charted the major features in a systematic way. To this day his alphabetical names are used for some of the lines, such as the sodium D lines. The absorption lines assumed very great importance for physicists when G. Kirchhoff and R. Bunsen (of Bunsen-burner fame) began to match patterns of absorption lines in the solar spectrum with the bright-line emission spectra of the atoms in hot gases observed in the laboratory. By these means they started systematic atomic physics and began to find out what elements are mixed together in the Sun's outer layers.

Dark lines arise when atoms absorb light. What happens is that the pure light from the photosphere encounters cool atoms. These can absorb energy at certain precise wavelengths, when the energy in the light precisely matches that needed to push an atomic electron out of one energy state and just into the next. When the electron tumbles down again it re-emits the energy, but in a totally random direction. Along a particular line of sight therefore light is depleted at the wavelengths associated with atomic transitions. Each element absorbs at a different set of wavelengths because the electron energy levels vary from element to element. The atomic spectrum of the atoms of an element is therefore unique to that element, providing an important way of seeing whether or not the element is present in a star's outer layers.

On the Sun most of the absorption is imprinted on the spectrum in a layer about 500 kilometres (300 miles) thick. It is not a uniform layer of cold gas: at its base a little absorption takes place, but there is still photospheric emission. Gradually the situation alters as lower temperatures are encountered at higher altitudes, so that at the top of the layer only absorption occurs.

Spectroscopic detective work has allowed astronomers to see what the sun's atmosphere is like in terms of composition, structure and velocities. The elements present have been tracked down by matching patterns of lines in the solar spectra with laboratory spectra. About sixty of the ninety-two naturally occurring elements are definitely seen in the solar atmosphere. There are other reasons, stemming from the chemical analysis of meteorites, for believing that a further two dozen elements are present. However these are not contributing detectable lines, either because their atomic structure is unable to make suitable transitions at solar temperatures or because there are only minute traces of the element. Among the missing elements from the periodic table are the group of related elements known as the actinides. These highly radioactive elements (polonium, astatine, radon, francium, actinium and protoactinium) result from the radioactive decay of the much longer-lived elements uranium and thorium. So their absence is entirely to be expected. Some of the elements are found in molecular rather than atomic form. Thus, the presence of the element fluorine is inferred from molecular transitions in magnesium fluoride and strontium fluoride.

One very striking result of early solar spectroscopy was the discovery of an entirely new element—helium, the second lightest gas. The English scientist Norman Lockyer tracked it down at an eclipse in 1868 and a quarter of a century passed before it could be studied in laboratories. A determination of the actual amount of helium present in the Sun is fiendishly difficult because the lines due to helium are very faint. Only at temperatures much higher than those encountered in the Sun can helium be stirred into action. What is certain is that helium is the most abundant element on the Sun, after hydrogen. The general opinion of solar astronomers is that helium makes up between 15 and 35 per cent of the atmospheric mass, with 25 per cent being a reasonable compromise value. To all intents and purposes the hydrogen content by mass is about three-quarters and helium one-quarter. All the other elements put together make up only 1.3 to 1.8 per cent of the atmosphere, and yet they are responsible for the appearance of over 20,000 Fraunhofer lines (not that Fraunhofer found all these, credit for that instead going largely to H. Rowland). Iron, which exists in the solar atmosphere in several distinct atomic arrangements, produces several thousand lines. Elements are detectable even though their abundance may be down by 1,000 million or more relative to hydrogen.

The finding of the relative composition of the various elements in a star's atmosphere from the Fraunhofer absorption lines is an extremely exacting business. First spectroscopic material of the highest quality is needed, it being important to spread the spectrum along the plate in order to make the subsequent measurements more precise. The information in the spectrum is then turned into a graphical or computer-readable form by measuring the varying intensity of a narrow beam of laser light as the photographic plate is moved across it at a steady rate. For particular elements, information is thus built up in the form of *line profiles*, plots of the intensity against wavelength in the neighbourhood of a line. In the Sun, hydrogen has a deep and broad profile whereas there are many thousands of sharp, shallow lines of iron. Turning these line profiles into quantitative data is a fine art.

The ease with which a spectral line can form varies very much from element to element. The line profile, which can be regarded as a signature containing information on the concentration of an element, is a function of the relative concentration, the temperature, the pressure, and the atomic parameters for the element in question. Now the effects of the electron energy levels can be calculated or observed in a laboratory; essentially this takes care of the fairly obvious fact that in the atoms of some elements (sodium, calcium and iron for example) the outer electrons are arranged in a way that makes absorption easy and likely, whereas others (helium is the classic example) find it very difficult. Once these matters of atomic structure are understood, the main quantities to take care of are temperature and abundance. The former can be found without much trouble; after all it has the same value for all the elements. This leaves abundance as the only major unknown quantity in spectroscopic work.

The solar or stellar spectroscopist still has to juggle with an enormous number of variables. For stars, much of the information is extracted by comparing the line profiles element by element. For the Sun, we can do a little better than this by constructing a *model atmosphere*. In essence this is a computer simulation of the solar atmosphere which solves the equations governing the transfer and transport of radiation in the Sun's cooler outer layer. Synthetic line profiles can be plotted for various physical conditions. By matching synthetic and observed line profiles it is possible to deduce chemical abundances. The solar reference atmosphere is now well explored by computer, but similar work on stars is not generally so far advanced.

The overall composition of our Sun's atmosphere is not unlike that of most other stars formed in the last few billion years. Compared to very old stars they contain between ten and a hundred times the quantities of elements heavier than hydrogen and helium. This fits in with the general idea that when the Universe was young the heavier elements were extremely scarce. The first supplies of heavy elements were manufactured by nuclear reactions in stellar and possibly galactic explosions. By the time the Sun was forming, however, the early generations of massive stars had already enriched the interstellar medium by conveniently exploding at the ends of their short lives.

With a couple of rather specialized exceptions, the composition of the heavy elements in the atmosphere of the Sun is the same as that in the gas cloud from which the Sun formed. The Sun is not manufacturing fresh supplies of heavy elements to enrich its own atmosphere. (I am excluding consideration of the possibility that the Sun's atmosphere has been contaminated with helium from the core through mixing. In any case the present Sun is not manufacturing anything heavier than helium.) The Earth and other planets also condensed from the same nebula as the Sun, so presumably solar spectroscopy also informs us on the make-up of the stuff from which the Earth formed.

The solar atmosphere contains traces of lithium and beryllium, two of the lightest elements, and this calls for an explanation because these elements are consumed in nuclear reactions. At temperatures as low as 1 million degrees the nuclei of these elements will capture colliding protons and then fragment into helium nuclei. As the proto-Sun formed it should have burnt up the lithium and beryllium in the earliest stages of its existence. Perhaps the atmospheric layers have never mixed to any appreciable extent with slightly deeper layers where the temperature exceeds a million degrees. This is a possibility, but it's a little hard to square with the foaming turbulence observed in the outer layers which must be causing some mixing. Observations made from the satellite OSO-7 in late 1972 pointed to a more intriguing explanation.

The gamma-ray detector on this satellite picked up two strong lines in the gamma-ray spectrum. One of these lines has the energy that photons acquire when an electron and a positron smack head on and annihilate, not in a puff of smoke, but into two photons with equal gamma-ray energies. The higher energy line matches the photons that are created when protons and neutrons crash together hard enough to stick and make deuterium. What the gamma-ray experiment illustrates is that in areas of great activity on the Sun nuclear reactions —that is to say the destruction of electrons and the fusion of proton-neutron pairs—take place. Nuclear reactions in the atmosphere offer a plausible explanation for the lithium and deuterium seen in the Sun. They are made in flaring and disturbed regions where protons are accelerated to very high energies. In certain rare stars, but not the Sun, spectroscopists find lines that betray technetium, a radioactive species with a half-life of only 200,000 years. This also must be manufactured in nuclear reactions in the atmospheres of the stars.

The Fraunhofer lines carry other information in addition to the presence and abundance of chemical elements. The *exact* wavelength at which a line is found and its width can be affected by magnetism, gas pressure and velocity along the observer's sightline. As far as velocity is concerned, the wavelength is affected by the well-known Doppler effect; that is to say, if the source of absorption is travelling towards us we see it at a shorter wavelength (shifted towards blue) whereas something moving away is at higher wavelength (shifted towards red). The amount of shift is proportional to the velocity for speeds significantly less than that of light.

Doppler-shift measurements can be used to find the speed at which the Sun rotates. The Sun does not rotate rigidly like the Earth. Being made entirely of gas it can swirl round faster at the equator than at the pole. The Doppler-shifts show that the Sun takes about thirty-seven days for one circuit in the polar regions. Moving to the equator, the gas dashes round faster and faster, taking only about twenty-six days at the equator. These times I have quoted are relative to the

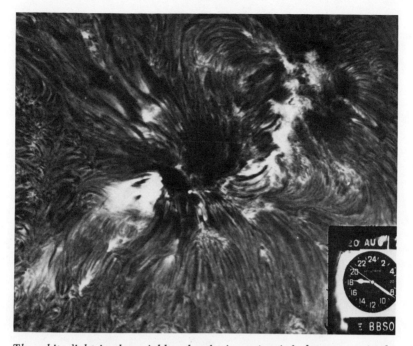

The white light in the neighbourhood of a pair of dark sunspots is the brilliant emission from a solar flare. When flaring occurs, nuclear particles may be accelerated to sufficiently high energies for a nuclear reaction. This mechanism may account for the gamma-ray line emission from the solar surface. (Big Bear Solar Observatory, California)

distant stars. We on Earth measure different rotation speeds because our planet is travelling round the Sun in the same sense as solar rotation. Consequently the rotation periods we measure, relative to Earth, are fourteen to twenty-seven days from pole to equator. If you compare these values with those in other sources they may seem slightly long. Most authorities quote rotation speeds based on the movement of sunspots. As we shall see these are severely influenced by magnetism, so they may not be entirely reliable as indicators of the 'real' rotation period. If the Sun does have a fast spinning core the magnetic field and sunspots will be speeded up.

Another important disclosure from the study of velocities by spectroscopy was the discovery, in 1960, that the solar atmosphere is

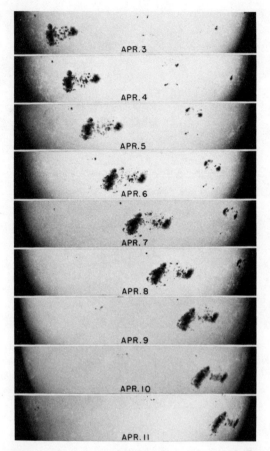

APR. 3

APR. 4

APR. 5

APR. 6

APR. 7

APR. 8

APR. 9

APR. 10

APR. 11

A classic sequence of sunspot photographs made in 1947 at the Mount Wilson Observatory shows the rotation of the Sun. The true speed of solar rotation is more reliably determined from spectroscopic measurements because sunspots are strongly affected by deep-seated magnetism. (Hale Observatories, California)

breathing in and out with a well-defined average period of five minutes. The velocities associated with this oscillation of the atmosphere are about 0.5 km per second (over 1,000 miles an hour). Gas immediately above the convection zone is moving up and down in unison, travelling about 50 km (30 miles) vertically in each oscillation. Imagine the weather on Earth if our atmosphere jumped up and down every five minutes or so! The layer of atmosphere above the convection zone proper has a natural resonance frequency for sound waves, just as an organ pipe has. Solar physicists conjecture that pressure waves, or sound waves, are produced in the convection zone at preferred frequencies that match the resonant frequencies of the layers immediately beneath the photosphere. In this way waves are fed into the atmosphere, causing it to bounce in and out.

I've already mentioned the oscillations found by Henry Hill; these are vibrations that penetrate to the heart of the Sun. On the other hand, the five-minute oscillation is basically a ringing tone in the outer layers. By marrying the two oscillation regimes together, solar astronomers have another means of poking about to see the Sun's insides.

A remarkably effective method of investigating happenings in the photosphere and atmosphere of the Sun is furnished by narrow-band filters. These can isolate the radiation being sent out by one particular element by transmitting a very narrow range of wavelengths centred on a spectral line. The filters, which exploit the interference of light to make sure that only the selected wavelength is passed with appreciable intensity, are an important diagnostic tool. This is because the temperature in the atmosphere varies with height. This in turn means that different atomic species are the dominant source of line absorption (and sometimes line emission) at differing depths. A narrow-band filter can be selected to isolate the radiation from particular layers. In this way the solar scientist can peel back the masks of the Sun.

Let's take isolating the chromosphere as our first example. The pinkish light from this layer, readily visible to the naked eye only during eclipses, mainly arises from emission in the first spectral line of the hydrogen Balmer series. This line has a wavelength of 656.3 nanometres. A filter which is only transparent to light between 656.25 and 656.35 nm will isolate this line. In the photosphere the 656.3 nm line is one of the darkest, or most *strongly absorbed*, features, so no light whatsoever from the yellow-white solar disc will pass through this particular filter. Instead only the light emitted by hydrogen in the chromosphere can get anywhere. Thus we have a method of photographing this particular solar layer despite the fact that it cannot be perceived purely visually except in an eclipse. In addition to the hydrogen line, a line of ionized calcium at 393.4 nm is frequently employed to explore the structure and activity of the chromosphere.

The pattern of brightness visible when the chromosphere is imaged through filters is called the chromospheric network. It matches the outlines of the deep-seated convection units known as supergranule cells and regions of higher magnetic field. Like the Earth's cloud cover the network continuously changes, having a characteristic lifetime of about an Earth day.

If, alternatively, the solar disc is imaged in the red light of hydrogen or the blue light of ionized calcium, a network is visible in the photosphere also. It coincides with the gross structure in the chromosphere. High-resolution photography, in conjunction with filters, re-

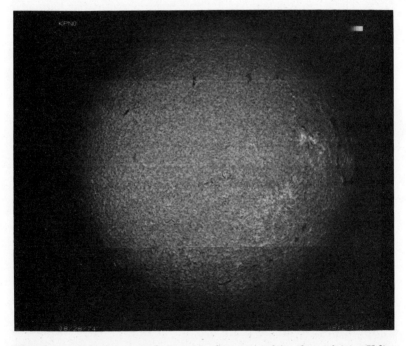

The chromospheric network is perfectly captured in this calcium K-line (393.4 nm) filtergram from the McMath solar telescope. (Kitt Peak National Observatory, Arizona)

veals a rich and varied structure—the pores, hairs and warts of the Sun's skin. Many fine dark lines, resembling a scattering of blades of grass or straw, are swept across the surface in whorls and circles. They are dark when seen against the face of the Sun, but bright at the limb, where they can be viewed against the blackness of space. Solar astronomers use many words to describe the Sun's dermatology; they call the dark lines *fibrils* and their bright counterparts at the limb *spicules*. These features are located in the low chromosphere and are found mainly at the boundaries of supergranules. Looking like jets, flames or burning hedgerows, each spicule lasts about two to ten minutes before being replaced by a new one. The Skylab mission produced many thousands of photographs of the spicules and the network structure in the upper chromosphere. This high chromospheric layer has a temperature of 70,000 K and is punctuated by bristling rows of spicules that reach 25,000 km above the surface. At the Sun's polar cap Skylab's ultraviolet imaging system revealed giant spicules darting 40,000 km high and twice as wide as the Earth. These macrospicules last nearly an hour. They are just one manifestation of the churning structure of the chromosphere, where matter is sloshed about at velocities of over 150 km per second.

I now want to describe an invisible component of the atmosphere —the solar magnetic field. The instrument known as a magnetograph plots the magnetism of the Sun. The physical principle employed in such an instrument is that magnetism distorts the electron structure in atoms. As a result their energy layers split to show a finer structure. Spectroscopically the individual lines are seen to split into patterns of closely spaced lines. Furthermore the light in the split lines is polarized. By a suitable combination of filters, polarized lines may be isolated and compared, so that the direction and intensity of the magnetic field may be deduced. Today this operation is carried out routinely by magnetographs that display the Sun's instantaneous magnetic field either electronically, or on film, or via television camera tubes and monitors.

Solar magnetism is formidably complicated. On the Earth matters are relatively simple: our planet has a permanent magnetic field which is dipolar—similar to a bar magnet—with two magnetic poles. The Earth's bar magnet drifts around and at intervals of a few hundred thousand years reverses in direction. But on a day-to-day or yearly basis not much happens. Geophysicists believe that the

A hydrogen alpha (636.30+ 0.08 nm) filtergram shows spicules at the edge of supergranule cells. Small bright mottles mark the base of the spicule rosettes near the photosphere. There is a small active region in the central foreground with absorption (black) at the top of the loops. (Sacramento Peak Observatory, Sunspot, New Mexico)

terrestrial magnetism is generated by a dynamo-type mechanism in the Earth's liquid metal core. The Sun's field has a great deal of structure and it changes continuously. Furthermore, the Sun's field flips over every eleven years or so, one manifestation of the semi-regular cycle of changes on the Sun.

If you could go very far away from the Sun and measure its magnetism by the method that we apply for any star, the result would be boring. Its magnetic-field strength would seldom be seen to exceed 1 gauss, about ten times stronger than the Earth's average field, and comparable to a very cheap toy magnet. Other stars, white dwarfs for example, possess intense magnetism measured in thousands of gauss. And in a neutron star or pulsar the surface magnetism reaches billions of gauss. Apparently the daytime star has little to offer the student of stellar magnetism.

But now let us move in closer and closer, resolving first the disc, and then large-scale features. In the polar region (above 60° latitude on the Sun) there is a well-defined pattern of lines emerging radially from the pole. This changes but slowly over the years, and is composed of many smaller elements of magnetism, rather than being a general magnetic field.

Skylab revealed the magnetism at the north and south poles in a glorious fashion. The new view from space and in ultraviolet light, was able to reveal the tracery of the field lines that eclipse photographs had merely hinted at. In the polar regions the field lines are open and reach out to space, consequently matter is able to flow out easily. This is the domain of the macrospicules, where vast volumes of the solar atmosphere are vertically stretched. The free flow of material causes gracious plumes, anchored to the bright spots in the corona, to stream into space.

High-resolution observations of the magnetism definitely show a profusion of jumbled knots, each with a field strength of 1,000 gauss or more, which almost cancel each other out, leaving a general field of a mere 1 gauss. Across the photosphere large-scale structure is evident, but again it is built from smaller elements. This magnetism stretches out into space, giving rise to a sectored structure in the interplanetary magnetic field. Within a sector one polarity predominates, and in the next sector the opposite polarity; these sectors can be traced back to the solar surface.

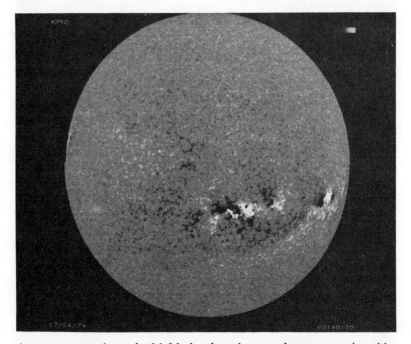

A magnetogram from the McMath solar telescope shows areas of positive (white) or negative (black) magnetic polarity. There is a large spot group at lower right that is about to break up. The strong surface magnetism is associated with solar active regions, which are areas of the Sun where the strong field of the interior wells up through the photosphere. (Kitt Peak National Observatory, Arizona)

High resolution will still reveal that in the photosphere much of the magnetism is concentrated into a filigree pattern that, like the spicules and fibrils, clusters round the boundaries of the supergranule cells. The strength of the solar magnetism inside the small magnetic elements, which measure 200–300 km across, is about 1,000 gauss. Most of the solar magnetism is concentrated into cells of this type.

The highest magnetic fields, of several thousand gauss, are associated with sunspots and regions of stormy activity. Sunspots have been known since antiquity, although the followers of Aristotle refused to believe in them. We know, from completely independent manuscripts in the Far East, that oriental scholars made thoroughly competent observations of sunspots hundreds of years before the invention of the optical telescope. They wrote picturesque descriptions of spots 'as large as a plum', or 'as big as a crow'. If the local atmosphere is transparent large sunspots can be seen without difficulty when the Sun is close to the horizon. The Aristotelians, however, would have no truck with this. For them matters celestial had to be the epitome of perfection. They reasoned that the Sun, being a sphere, the perfect shape, could not possibly be pockmarked with ugly black spots.

This attitude, transmitted through to European thought by Thomas Aquinas, was to cause Galileo a good deal of trouble. Christoph Scheiner, a Bavarian, made small telescopes and used these to observe the Sun by the projection technique. After studying the Sun for seven months he realized that spots disfigured the brilliant image. Several observers were able to see the spots, which suggested that they could not be an optical illusion. Furthermore, they had the same appearance in different telescopes and so could not be artefacts caused by lens imperfections. Scheiner also ruled out clouds high in the Earth's atmosphere as an explanation, because the spots did not change their position on the Sun when observed from different locations. No, he concluded, they had to be on or very near the Sun. He eventually decided that they were planets or something similar orbiting above the solar disc, reaching this conclusion largely because it allowed the Sun itself to remain pure, in line with the traditional philosophical viewpoint.

Galileo got to hear of Scheiner's discoveries, and set to work. He had no illusions about the Sun, which he regarded as an imperfect object anyway. Being a cunning political operator, Galileo claimed priority of discovery, asserting that he had been watching sunspots for well over a year. Be that as it may, he certainly sketched and recorded the spots, and made many fundamental findings. He observed that they appear and disappear and also that they change in size; by watching them when close to the limb of the Sun he perceived the changes of shape caused by the foreshortening effect on a globe. Galileo was quite clear about sunspots: he stated that they were located on the Sun and that the Sun was a sphere. His inept and unfounded criticisms of Scheiner's work, however, stimulated the latter to bear a grudge and, later, to hatch plots against Galileo.

The apparently dark interior of a large sunspot is called the **umbra** and the fibrous greyish surround, the **penumbra**. The spots only appear to be dark because they are embedded in the brilliant photosphere. If a spot could be isolated it would have a surface brightness greater than an arc light, because the temperature is 4,000 degrees, about 2,000 degrees cooler than the photosphere. The average sunspot would appear as bright as the full moon if it could be seen against black sky.

William Herschel thought that sunspots must be actual holes in the Sun's fiery atmosphere. The dark interior was the surface of the

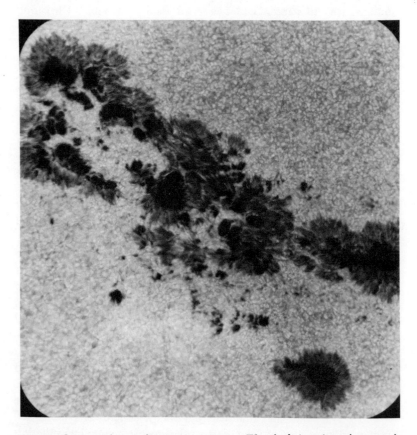

A magnificent and complex sunspot group. The dark interior of a spot is called the umbra and the surrounding region with a fibrous appearance the penumbra. In this case the solar granulation is also apparent. (Sacramento Peak Observatory, Sunspot, New Mexico)

inhabited planet beneath the Sun's fires. In 1795 Herschel wrote thus about the habitability of the daytime star:

> *The Sun appears to be nothing else than a very eminent, large and lucid planet, evidently the first, or in strictness of speaking, the only primary one of our system; all others being truly secondary to it. Its similarity to the other globes of the solar system with regard to its solidity, its atmosphere, and its diversified surface; the rotation upon its axis, the fall of heavy bodies, leads us on to suppose that it is most probably also inhabited, like the rest of the planets, by beings whose organs are adapted to the peculiar circumstances of that vast globe.*

The fact that sunspots are depressions has been known for some twenty years as a result of observations made by A. Wilson of Glasgow. He watched the progressive change of shape when symmetrical spots approached the limb of the Sun, and deduced that the spots are saucer-shaped depressions, because the dark umbra disappears close to the limb. From the effect of perspective as the spots approach the limb, scientists have estimated that the average depth from the photosphere to the umbra is about 700 kilometres (450 miles).

The dark umbral region covers about one-fifth of the area of an average spot. Activity is discernible inside the umbra. For example, under the best observing conditions it's possible to detect bright dots inside an umbra, usually about 100 kilometres in diameter. They survive for only a few minutes at a time. Modest flashes of energy, perhaps caused by magnetic waves bouncing around in the lower levels of the umbra, are also seen. A spot may also act as a resonant cavity, because vibrations of the umbra have been detected as well.

The grey penumbra looks like a striation or radial network of bright filaments on a dark background. Even a small telescope will show this structure. The way to follow the sunspots and their development, a fascinating project, is to use the projection method explained on p. 33 for a few successive days.

It was studies of the sunspots that revealed the fact that the Sun rotates more rapidly at the equator than at the poles. If you watch sunspots disappear round the limb of the Sun you may see them reappear at the opposite limb about fourteen days later. The average rotation speed for sunspots, as viewed from Earth, is 27.28 days, those at the equator going a little faster than average and those at latitudes 40° from the equator a little slower.

Where do spots come from? This is a difficult question to answer in physical terms and it is taken up in the next chapter. What you actually first see is the development of dark pores in disturbed regions of the photosphere. Pores or tiny sunspots typically (but not always) form into a pair of sunspots. The pores themselves just appear in about an hour and vanish after a day or so if they do not blossom into real spots. Nobody really knows why pores erupt when and where they do. The pairs lie along the direction of the Sun's equator. The leading one (that's the one that leads, relative to the rotation, and will reach the limb first) is usually more compact and it will move a little faster than the following part. It's not unusual for smaller spots to break through in the gap between the two main spots; if this happens the following spot may disappear, leaving just the leader, which itself slowly subsides. Spots come in many shapes and sizes. A great many of them are larger than the Earth, and rarely, perhaps once in a decade or less, a really large and complex group is seen splashed across one-fifth of the disc. Spots spanning more than 40,000 kilometres (25,000 miles) can be seen with the naked eye, but not without danger, when the Sun is very close to the horizon. The lifetime of a spot group can be anything from a few days to many months. In 1946–7 the largest spots to occur since the invention of the telescope appeared, and just one of these giants blotted out as much as 10,000 million square kilometres of the photosphere.

Associated with the spots is another type of wrinkle in the Sun's battered skin: the faculae. These are regions of great brightness, more than the usual photosphere, which look a bit like mountains of gas straggling above the photosphere. Bright and well-structured faculae quite often surround sunspots. Photographs taken through narrow-band filters can separate the faculae from the disc. The faculae can be followed through all layers of the solar atmosphere, fanning out and generally becoming more diffuse higher up. As far as photography is concerned, it's often much easier to see them near the limb as the perspective accentuates them. Identifiable faculae do not last anything like as long as spots, just a few hours being the average performance.

Space-age astronomy revealed many new facts about the sunspots and related blisters. One of these is that the direct influence of sunspots extends up through the atmosphere and into the corona. Plumes of gas arching between sunspots are seen; these are visible because the gas in them is cooler (300,000 degrees) than the crystal-clear, extremely hot (2 million degrees) corona.

There is so much going on at the photosphere and beneath the skin that it will be a long time before we understand fully the surface layers of the daytime star. Here physicists and astronomers can daily witness a fascinating struggle between the pressures of hot gas and tangled magnetism, as the prolific solar radiation battles to launch its way to the rest of the Universe, flaming along at the speed of light. Gas with temperatures of hundreds of thousands of degrees flies up into and glides down from the corona. Dark spots wax and wane, making holes 800 kilometres deep. This is a physicist's paradise, watching the forces of nature play each other off in a 10-billion-year battle that only gravity can win.

Activity and the Solar Cycle ⑨

Sunspots are just one example of the 'weather' in the photosphere and atmosphere. The active Sun also has flares, prominences—giant tongues of gas leaping into the corona—and bursts of radio noise. These are examples of energy release that occur when the outer layers of the Sun get mildly unbalanced, frequently because the magnetic field is getting into a frightful twist. The wiggly nature of the magnetic field is the key to an understanding of violent energy bursts on the Sun. Unlike planet Earth, the daytime star keeps fiddling about with the magnetic field, winding it tighter and tighter, until in the end it snaps like an over-wound clock spring. The term *solar cycle* refers to the almost regular series of observed changes on the Sun that accompany the winding up and subsequent release of the magnetic field. This intriguing pattern of behaviour first came to the attention of astronomers through observations of sunspots.

For nearly three centuries now, astronomers have made worthwhile observations of sunspots, and there is rudimentary information going back even further. Galileo made his observations in 1610–11, and records were kept on and off from that period. By 1843 sufficient data had accumulated for H. Schwabe of Dessau to confirm the long-held suspicion that the sunspot activity fluctuates regularly. Schwabe showed that the spots run through a cycle in which their number reaches a peak about every eleven years. The next person to contribute materially to sunspot research was R. Wolf, who, in the mid-nineteenth century, gathered all the data he could lay his hands on and reduced it to a convenient form. He established an average period of 11.1 years.

To make sense of the subjective judgement involved in deciding how spotty the Sun is, Wolf's own definition of sunspot number is still used. This number, a measure of solar acne, takes into account the number of spot groups as well as the number of spots seen on a given day. Each group gets a score of ten points and each spot scores one.

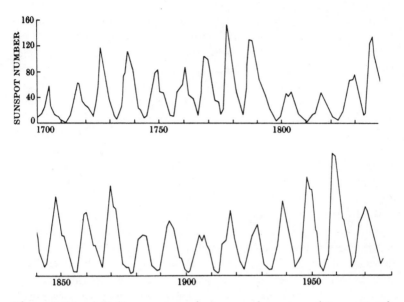

The variation of Wolf sunspot numbers over a long span of time not only shows the 11–year solar cycle plainly but also hints at the possible presence of a longer cycle having a period of around 80 years.

The total score for the day is the Wolf sunspot number; it can be as small as zero or as large as 200. Now, how did Wolf decide to give a score of ten for a group? He had to choose something, that's all; there is no physical basis, but the scheme is rational, giving a high weighting to the extent to which spots are clustered. The proof of the pudding is in the eating—astronomers still use Wolf's system because it works. One last point needs mentioning—there is a system for adjusting the score (a bit like a handicap in a horse-race or a round of golf) to take account of variations in observer enthusiasm, equipment and the weather.

A plot of the monthly average sunspot number shows with great clarity the cyclic behaviour of sunspot activity. In the last half-century the cycle has speeded up (but perhaps not significantly) and come down to around 10.5 years. The 200-year average gives a period of 11.2 years. Over the last 300 years it has been as short as 7 years or as long as 17. In other words the behaviour is semi-regular. If you look at the variation in sunspot numbers over three centuries there seems to be a pattern in the rise and fall of the peak. Perhaps there is another cycle of around eighty years modulating the eleven-year one—we won't really know for another few hundred years! Notice also that the rise to the top of the peak takes less time, perhaps four years, than the fall, which generally takes about six years.

Although Wolf's scoring system has withstood the test of time well, it is today more sensible to measure spot activity quantitatively. This is just what the observatories who keep regular patrols now do, by estimating the sunspot areas in units of one-millionth of the area of the Sun's visible hemisphere.

While the sunspot numbers increase, the spots also migrate towards the solar equator, which, incidentally is tilted at an angle 7° to the Earth's orbit (i.e. the ecliptic). G. Spörer (1822–95) of the Potsdam observatory was the first to investigate the variation with latitude. He and Richard Carrington (1826–75), an English amateur astronomer who made many contributions to solar astronomy, made a long series of observations of the rotation periods of the spots. From these measurements they confirmed that the Sun does not rotate as a rigid body. They found that spots at latitude 30°, for example, took about 7 per cent longer than equatorial spots did to make one circuit round the Sun.

E. Walter Maunder, superintendent of the Royal Observatory at Greenwich, London, published an entertaining diagram in 1904 showing the migration of sunspots with respect to latitude during the solar cycle. This *butterfly diagram* very clearly shows the steady shift from higher to lower latitudes. In general the first spots of a new cycle appear at about latitudes ±30° on the Sun, although they may be as far out as ±40°. As the cycle proceeds the spots are seen to occur progressively further in. They have reached ±15° at the peak of the sunspot number, and become clustered near the ±5° latitude at the end of the cycle. Note that individual spots do not move towards the equator; only the place where new spots are most likely to occur actually migrates. At the end of a cycle there may be spots from the old regime down near the equator while new ones are erupting in higher latitudes.

George Ellery Hale detected sunspot magnetism in 1908. In the 1950s the Babcocks, using a more effective magnetograph, showed that solar active regions have bipolar magnetic fields. The behaviour of the magnetism inside sunspots gives us clues as to the nature of the engine that drives solar activity. Magnetic-field measurements in pairs of sunspots clearly show that the two spots in a pair have opposite polarities, indicating that the magnetic-field lines emerge at one spot and re-enter at the other. During a particular solar cycle, and in a given hemisphere, the leading spot (leading in the direction of solar rotation) is always of the same polarity. In the hemisphere on the other side of the solar equation the same is true but the sense of the polarity is switched. This behaviour persists throughout the solar cycle and then when the next cycle gets into gear switches over

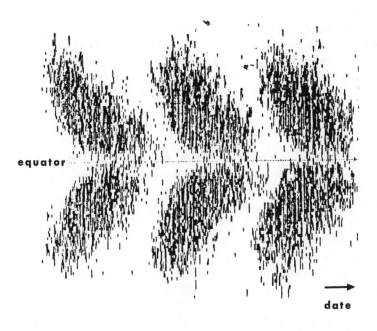

equator

date

Maunder plotted a diagram to show the latitude of sunspots as a function of date. The characteristic butterfly diagram results as the spot location moves progressively to the equator. An individual spot does not move, only the average latitude in which spots are found.

completely. The end of one sunspot cycle and the beginning of the subsequent one is accompanied by a switch-over of solar magnetism during which even the weak general field slowly reverses direction, taking about a year to do so. So the *magnetic cycle* of the Sun is a full twenty-two years, give or take a few months.

Although sunspots are the sole sign of action visible to the unaided eye, there is a lot more to solar activity than these blemishes. An active region on the Sun, a highly disturbed zone often, but not exclusively, associated with sunspots, is the stamping ground for many phenomena: faculae, flares and prominences, to name a few. The one linking theme of these antics is the intense magnetic field, measuring thousands of gauss. In fact, the active regions, whose size ranges from 10,000 up to perhaps 500,000 kilometres from end to end, provide the most striking evidence of the magnetic life of the Sun.

Faculae are regions in the photospheric layers that appear bright in white-light photographs and filter photographs. Near to young and healthy active regions the faculae are dense and bright. Hevelius, in the seventeenth century, seems to have been the first sun-watcher to notice faculae. Closely related to the faculae are plage regions; these occur in the chromosphere and merge with the inner corona. Emission of energy from the active regions can be mapped at optical, ultraviolet and X-ray energies, one of the important applications of orbiting observatories being to get the ultraviolet and X-ray pictures. At higher photon energies the active region sprawls out and has a less sharp structure, yet is still constrained by the magnetic field.

It is in the raging chromosphere that many of the effects of solar activity are most keenly felt. This region and the unstable layers above it are not easily studied from the ground, so an important task for the Skylab mission and other satellite programmes was to probe solar activity. Many of the events are best studied at ultraviolet and X-ray energies because their violent progress is mainly advertised at those wavelengths. Sunspots and large active regions were easy targets for Skylab. Transient and explosive events needed careful watching in the craft and at ground-based observatories. This co-operation has led to a much clearer picture of unrest in the chromosphere and corona.

From the sunspot data and information about active regions a model for the activity can be pieced together, but we stress that this is a *model* and not a formal theory. Really what we are trying to do is explain the weather machine in the Sun's outer layers, just as meteorologists account for the depressions, highs and fronts in the Earth's weather. In terms of theory and prediction the meteorologists are much further forward than solar astronomers. This is hardly surprising since mankind has a much greater incentive to understand the weather on spaceship Earth before grappling with the Sun. And, more seriously, the solar weather is to do with hot plasma, which is very slippery stuff, squirming through magnetic fields, which leads to problems of great mathematical complexity. On the other hand, deficiencies in our understanding of the activity do not have serious implications as regards the study of the Sun as a star, because the most violent storms only change its energy output by a factor of one-millionth, and this cannot change the evolutionary pattern.

It is widely accepted that solar activity is magnetic in origin and arises because the Sun is not rotating as a rigid body. We see the equatorial zone constantly lapping the polar regions, and it is possible, but neither proven nor disproven, that the insides may be spinning faster than the outsides. So where does this take us?

First we must firmly establish one important idea; this is that the Sun's gases are hot, and as a result there are plenty of free electrons, resulting from the partial break-up of the electroh clouds in the atoms that find the going too hot. Electrons carry the electric current; matter which contains free electrons—like iron and copper—is a good conductor of electricity. (The free electrons are found in cold metals because the atomic arrangement of the crystal lattice causes a separation into positively charged atoms and spare electrons.) When electrically charged matter that has a high conductivity tries to move against a magnetic field it finds it cannot manage it. As it tries to do so, the magnetism causes electric currents to flow in a way that builds up a secondary magnetic force opposing the motion. If that all sounds confusing here's another way of looking at it: when a cloud of electrons is moved relative to a magnetic field they automatically generate a further magnetic field that opposes rather than accelerates the motion. If this were not the case perpetual motion machines would by now provide all motive power and energy needed on Earth.

What happens on the Sun, then, is that the solar magnetic field gets trapped, or frozen-in, to the hot gas. As the Sun's gases glide around they take as much of the magnetism with them as they can manage. Since the equator overtakes the poles the lines of force get stretched;

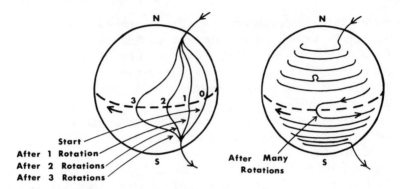

The differential rotation of the Sun gradually winds up the lines of magnetic force.

unlike spaghetti the field lines don't snap when they are stirred—they are more akin to super-stretchy elastic. The more they are stretched the more energy is stored in them, just like elastic.

Now imagine starting with a simple Sun: neat dipole field, just like a bar magnet, with undisturbed field lines connecting the poles. Next make it rotate, with the equatorial gas going faster than that at higher latitudes. After a few dozen rotations, the initially simple field lines have become wrapped around the Sun a few times. This process continues, and each time the equator laps the pole the magnetic vice encircling the Sun is tightened as the field lines get squashed closer together. Furthermore, what was once a magnetic dipole field is gradually changed to an intense ring-doughnut (or torus) field. The field lines jostle for room. Eventually something must give way.

When the field strength in any part of the outer layers reaches around 10,000 gauss (this is about 100,000 times the Earth's field), the magnetic repulsion is fierce enough to balance the inward pull of solar gravity. Now the plasma writhes and twists, tangling the field lines even more, as convection stirs the upper layers. The field is jumbled into ropes and knots. In places it breaks through the photosphere, forming the emerging flux regions, which are the first stage in the creation of a solar active region. New magnetic field lines are thus heaved up to the surface of the Sun. The regions where they break through have the bipolar structure—pairs of north and south magnetic poles—first seen in sunspots in the early twentieth century. This stage may be accompanied by a bright plage region. About a day or so later the sunspot pair itself emerges, and an arching structure of filaments, which presumably mimics the structure of the magnetic field, connects the two spots. These arch filaments can be 30,000 km (20,000 miles) in length and 5,000 km (3,500 miles) high—in other words, much larger than the Earth.

Within the sunspot zone a tube-like magnetic field emerges from one spot and arches over to re-enter the Sun at the other. This picture gives a natural explanation of the dual polarity as well as being consistent with the observation of arch filaments. The spacecraft observations, of plage regions particularly, now allow this structure to be traced far above the photosphere.

In the late 1960s Spencer R. Weart made a detailed investigation of the emergence of new active regions on the Sun. By use of time-lapse movies of the Sun it was possible to follow the growth of a sunspot region back in time until the first few hours of its appearance. One surprising finding was that the arch of magnetism that is forced out of the Sun is initially at a random angle with respect to the equator. Yet, within a few hours the emerging tube gets twisted into place by magnetic forces and the reorganization of field lines. In this way the Sun ensures that the sunspot pairs will be parallel to the equator. The fact that the orientation is almost random in the earliest stages suggests that the magnetism beneath the surface may be a chaotic jumble of field lines.

When the arch of magnetism for a new spot group emerges it gets twisted into place. The asymmetric nature of sunspot regions may arise in the following way. If the emerging magnetism only needs a small twist, the sunspot region grows. If, on the other hand, it has to be wrenched through a large angle, the 'wrong' spot is the larger and the group soon fades away. For this reason, in almost all spots actually observed, the preceding spot is larger; the other groups simply do not survive.

Skylab gave astronomers their first opportunity to study the extensive structures of active regions that extend into the upper chromosphere and lower corona. Observations made simultaneously at several of the 'invisible' wavelengths allowed mapping of the delicate tracery above the photospheric active regions. A most important finding was that the active regions are definitely governed by arched magnetic tubes. High-temperature gas is trapped within these tubes anchored to the face of the Sun.

At solar maximum the new regions pop out of the photosphere at the rate of one per day, buoyed up by convection in the middle of supergranular cells.

Solar physicists are still intrigued by the relatively cool interiors of the spots. Without doubt, a giant refrigeration process is at work, carrying heat away from the spot so efficiently that the temperature drops by almost 2,000 degrees. There are several possible explanations for this. One is that the strong magnetic field, which is locked in with the tumbling gas below the photosphere, can slow down convection a great deal, and so form an insulating layer under the spot, drastically reducing the energy flow. This is how fibreglass or mineral-wool roof insulation works, by cutting down the turnover of large convection cells just beneath the roof floor. Winter snow melts most slowly on the best-insulated roofs because they are the coldest. Another possibility is that above the spots material flows away speedily, expanding as it does so, because it swims down the field lines. This expansion of the plasma would cool down the photosphere locally. Finally, there is an idea that magnetic waves are generated over the spot, and that these can pump energy up to the corona. Theories, theories, theories, but which is correct? Maybe a combination—they all have difficulties, one of which is that the energy removed from the spot is apparently absent elsewhere. If the insulating-blanket model is right then why do we not see bright rings around the edge of the spot, where the diverted heat should be escaping? And, similarly, if some pumping mechanism is at work we might expect to see bright points in the corona. Refrigerators and freezers, after all, have to have a heat exchanger or fan to get rid of the waste heat.

During a solar cycle the new spots first blister in latitudes ±40°, which is where solar scientists think that the shearing of the magnetic structure first takes place. Magnetic buoyancy lifts the field from the lower convective zone 200,000 kilometres down. Eruptions then have the effect of taking the pressure off in the higher latitudes. However, differential rotation continues to wind up the lines nearer the equator, with the result that the spots gradually close in. The final stage of a cycle is marked when the field lines at the equator become so crowded that they short-circuit, cancelling out most of the magnetism. The new cycle begins with the field lines running in the opposite sense, pole to pole, because the dynamo field of the Sun turns over at the end of a cycle. When the first spots of a new cycle emerge they also have polarities arranged in the opposite sense to that of the previous cycle.

There are also considerable difficulties in explaining how the solar dynamo works, and the solution of this problem will include another imponderable—is the Sun rotating faster inside? For a time, in the early 1970s, it seemed that the Sun did have a rapidly rotating core, but this notion was called into question by the failure to detect the squashiness (oblateness) that it ought to cause.

It might be easier to solve the problems of the Sun's magnetic variations if they were regular, but even this has been brought into question by the fascinating results that have emerged from a diligent survey of historical records. These show that sunspots have not always come and gone in quite the way that they have in the last 250 years. In 1976, astronomer John Eddy reviewed the records relevant to sunspots over the past 1,000 years and concluded that the Sun has undergone significant change, meriting the closest scrutiny.

The clues to the existence of real changes came about in the following way. Galileo and Scheiner had observed them in 1610–11, but 230 years elapsed before Heinrich Schwabe found the apparent period separating maxima; at first sight such a long time to discover the solar cycle is scarcely a credit to the early solar astronomers. In the late nineteenth century two observers, Gustav Spörer in Germany and E. W. Maunder of England's Greenwich Observatory, pointed out in five research papers that in a seventy-year period up to 1716 there seems to have been a genuine near-absence of spots on the Sun's disc. For nearly half this time, that is from 1672 until 1704, practically no spots whatever were seen in the Sun's northern hemisphere. As far as sunspot groups were concerned, only one was noted in the sixty years preceding 1705. Maunder had the benefit of following Spörer in much of this research and was able to support the assertions by reference to the scientific literature of the period. Eddy, in his reassessment of the work, quotes a paper published by the Royal Society of London. This reported the sighting in the year 1671 of a sunspot: '. . . at Paris the Excellent Signior Cassini hath lately detected again Spots in the Sun, of which none have been seen these many years that we know of.' Cassini wrote that his discovery came twenty years after the time astronomers had last seen considerable spots on the Sun. And a final contemporary chord was struck in 1684 by England's Astronomer Royal, Flamsteed, who reported the appearance of a spot thus: 'these appearances, however frequent in the days of Scheiner and Galileo, have been so rare of late that this is the only one that I have seen in his face since December 1676.' In fact, by the time Maunder took up the question there were plenty of mentions in the literature of the missing sunspots.

In his analysis Maunder leaned heavily on a thin veneer of archival evidence, and, more questionably, on the specious argument that absence of evidence is evidence of absence. Jack Eddy has made a new assessment, taking account of many facts not considered previously. He came to the conclusion that Maunder's minimum was a real occurrence, not due to faulty or incomplete observations. For example, it has been found that records of displays of the aurorae (northern lights) suddenly increase in the early eighteenth century, when the spots returned; astronomers have now established that the auroral displays are more brilliant when there are many spots around.

Modern confirmation of long lapses of solar activity also comes from studies of the past abundance of the heavy radioactive form of carbon known as carbon-14. This isotope is a constituent of the carbon dioxide in the Earth's atmosphere which gets incorporated into the woody tissues of plants and trees. When the Sun's weather is calm and its magnetic field quiet, charged particles called cosmic rays, which are travelling all the time through the Milky Way Galaxy, are able to reach the Earth more abundantly. If the Sun is magnetically active it has many sunspots, and the increased magnetism also shields the Earth somewhat from galactic cosmic rays. The cosmic rays that do zip through our atmosphere manufacture carbon-14 as they collide with other atoms in the air. The net result then is that fewer-than-average (less solar activity and lower magnetism) will match more-than-average carbon-14 (because cosmic rays more readily collide with our atmosphere). Essentially, by measuring the carbon-14 in datable tree rings, scientists have found how its natural abundance has fluctuated in the past. When these studies first got underway scientists puzzled over the prolonged increase in carbon-14 between 1650 and 1700. Now it can be seen that this anomaly is closely related to Maunder's sunspot minimum.

Confirming evidence for the Maunder minimum, as well as strong indications of two earlier lows in sunspot activity, has come from study of the written histories and astronomical treatises in the orient. This work has been particularly consolidated by two astronomers in England, David Clark and Richard Stephenson. Although European sources have almost no mention of sunspots before the time of Galileo, the same is not the case in the orient, where a goodly sprinkling of sightings has been handed down to us. Why the difference? In Europe intellectuals believed that the Sun was perfect, and therefore the merest possibility of sunspots was ruled out on theoretical (or, more correctly, dogmatic) grounds. In the East this stranglehold on original thought did not prevail; hence there is an abundance of beautiful and poetic references to naked-eye sunspots. These examples are taken from the translation of Clark and Stephenson. 'The Sun was a dazzling red, like fire. Within it there was a three-legged crow. Its shape was seen sharp and clear. After five days it ceased' (AD 352). 'When the Sun first rose and when it was about to set, within it on both occasions there was a black colour as large as a hen's egg, and after four days it was extinguished' (AD 579). This is a fine example of how naked-eye spots are relatively more easy to see when the Sun is close to the horizon shining through a murky haze. Oriental descriptions over a 1,500-year period have many picturesque estimates of size: 'like a plum', 'as large as a date', 'as large as a coin' (28 BC), or 'black vapour like a flying magpie'. These estimates do not convey any quantitative estimate of size; however, the flying magpie group of AD 188 lasted through several solar rotations ('. . . after several months it gradually faded away'), so it must have been a whopper.

The oriental records seem to show two amazing gaps lasting about 200 years. One lasted from AD 600 to 800, when not a single naked-eye sunspot was recorded, and the other from AD 1400 to 1600 when the score was a mere two sightings. Interestingly, the twenty years preceding AD 1400 include a wealth of sunspot reports, showing that solar astronomy was flourishing. Apart from these two conspicuous breaks there are three smaller ones that could, conceivably, be due to a lack of observer enthusiasm rather than real absence of spots. The truly outstanding feature of the large gaps, however, is the coincidence with corresponding peaks in the amount of carbon-14 present in the atmosphere. A further gap, from AD 1280 to 1350, also follows an increase in carbon-14; this excursion from normal is called the *medieval minor minimum*. The gap from AD 1400 to 1600 is the *Spörer minimum*, and that following the invention of the telescope is termed the *Maunder minimum*.

The finding of long sunspot-free periods over nearly two millennia of solar observations shatters the conviction that the daytime star has a regular 11-year cycle. Clearly there is another profound effect at work, one that can switch off the spots and reduce the magnetic field. This phenomenon is confirmed by studies of carbon-14 in the remains of plants and especially in tree-rings. Further support comes from more indirect observations: the extent of the corona at solar eclipse is lessened when there is little action. During the Maunder minimum (whose reality cannot now be doubted) there was also a marked absence of aurorae, which we now know to be an indicator of violent magnetic storms on the Sun. All the signs are that the Sun is not the predictable variable star that astronomers have grown accustomed to, but rather a star that undergoes significant *unpredictable* changes in behaviour. As Jack Eddy has suggested, the Sun may currently be heading for a grand maximum in the twenty-second or twenty-third century.

The study of ancient sunspot records will clearly continue to be important. At present we do not know why the solar cycle is irregular, as it obviously is, nor do we have really firm ideas as to the effect the variations may be having on the strength of the Sun's rays and hence the effect, if any, on the Earth's weather.

The size of the corona as perceived at eclipse varies with the solar cycle, being relatively compact and uniform at minimum, but much larger and showing complex structure at solar maximum. When the Sun has many spots the corona is characterized by numerous long streamers that look like flower petals. The corona is also a good deal brighter at maximum. During the Maunder minimum, observers described the corona as a very limited dull glow. But only a few years later, in 1715, an observer at Cambridge gave the first reasonable description of the corona together with the streamers. At solar maximum the corona is, of course, breathtakingly beautiful. At minimum the true corona may even fade completely, the residual ring of light being sunlight scattered from dust in interplanetary space. The particle density in the corona increases twofold and the temperature by around 20 per cent between solar minimum and maximum.

Eclipses provide casual or, in the days of television, even armchair observers with an opportunity to see not only the corona but also prominences. These beautiful structures come in several varieties, only some of which are directly associated with the active Sun. Sun-

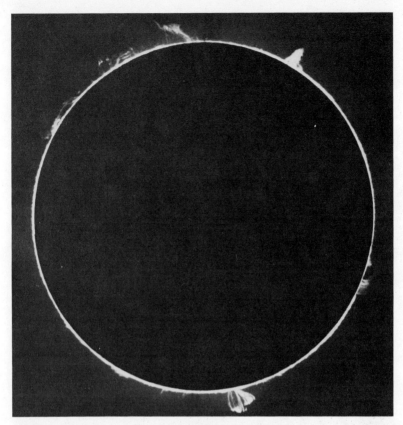

Prominences photographed at the limb of the Sun on 9 December 1929. (Hale Observatories)

watchers have known about prominences for centuries. Back in 1239 'a burning hole' was seen in the corona during an eclipse; in all probability this was a giant prominence. Apparently they were again reported by chroniclers in Russia during the Middle Ages. At the eclipse of 2 May 1733, B. Vassenius saw three or four prominences from Gothenburg, Sweden, which he called red flames; he believed them to be clouds in the Moon's atmosphere. Curiously all these observations came to be forgotten, so astronomers had a surprise at the eclipse of 8 July 1842 when several of them rediscovered prominences, which they interpreted as mountains on the Sun.

By the late nineteenth century, fortunately, real progress could be made, with the introduction of photography and spectroscopy. Spectra obtained at the 1868 eclipses showed brilliant emission lines in the prominences, and from then onwards they were correctly inter-preted as glowing clouds of gas, high above the surface of the Sun. Incidentally, it was at this eclipse that they discovered a bright spec-tral line that was not attributable to emission by any known atom, so it became associated with a solar element—helium.

What, then, is a prominence? The simplest, and not particularly scientific, description is that they are flame-like or curtain-like struc-tures when they are observed above the solar limb, as at eclipses. However, not all such structures are prominences—some are flares, of which more later. A slightly more scientific approach is to state that prominences are cool and dense masses of gas in the hot corona. They take on many different forms, lasting from a matter of months down to a few hours. The early solar physicists thought of them as violent eruptions—firecracker displays—of matter belched from the photosphere, but modern time-lapse photography shows that in many of them cool gas is steadily raining down through the corona onto the photosphere.

There are two major compartments that solar physicists use for filing personal data on prominences: one for *active prominences* and another for *quiescent prominences*, a division that dates from the year 1875. The active prominences have names such as surges, sprays, loops, and eruptive prominences, words that evoke the large-scale movement of gas at high speed and with violence. A brief survey of just some of the prominence types will give an idea of the range of types.

Coronal clouds hang in the corona, pouring gas into active regions in the photosphere beneath. They typically last for a day or two and are found at altitudes of a few tens of thousands of kilometres. A coronal-cloud prominence is generally rather larger than the Earth. With coronal clouds there may be a shower of coronal rain, this being visible matter that is streaming along curved magnetic field lines back into the photosphere and its active regions. Coronal raindrops splat-ter down at speeds of 50–100 km per second (100,000 miles an hour). Not all prominences shoot high into the corona; mound prom-inences, which are most readily seen near to the limb of the Sun, are low-lying features.

The spectacular prominences are of several types. In the tornados the magnetic field is a vertical spiral, causing the prominence to look like a twister. Loop prominences are formed of material which arches through the corona, being firmly anchored at each end in or near to sunspots. They are associated with the highest degree of solar activity. At the top of the loop the corona is extremely hot, and material cascades down through the two legs. Apparently explosive phenomena sometimes occur. Sprays are vigorous ejections of hot gas at speeds of 400 km per second (approaching 1 million miles an hour), in which the initial acceleration is tremendously high—a g-force of several hundred is not unusual. As this exceeds the minimum speed for material to escape from the Sun, some of it does indeed escape forever. Observations suggest that some of the sprays occur when a bubble of plasma tightly enclosed in a magnetic knot suddenly bursts its cage apart. Sometimes the material is fired out along a vertically ascending path in what is known as a surge prominence. These take about fifteen minutes to zoom 100 million kilometres into the corona and then they start to fall back along the original trajectory. Again, these prominences seem to be initiated by a bomb-like explosion of magnetic field and hot gas in the photosphere. The advent of orbiting space laboratories has, of course, allowed solar physicists to explore the properties of prominences and other ejection-phenomena at the

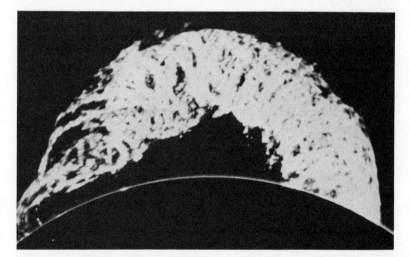

The great eruptive prominence of 4 June 1946, one of the largest ever seen. This prominence became as large as the Sun in only one hour, and within a few hours it had completely vanished. (High Altitude Observatory and National Center for Atmospheric Research, Boulder, Colorado)

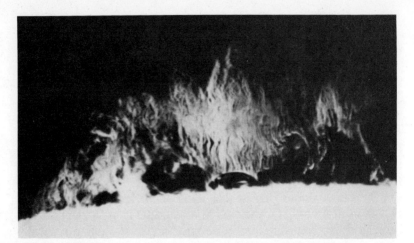

A small limb prominence about 50,000 km high. This shows vertical structure, which remains static while material flows down through the tubes moulded by the magnetic field. (Sacramento Peak Observatory, Sunspot, New Mexico)

invisible but energetic frequencies of the ultraviolet and X-ray electromagnetic spectrum. The ultraviolet spectroheliograph on Skylab recorded a spectacular solar eruption flashing out half a solar radius (fifty times the Earth's diameter). This turned out to be a jet of helium gas at a low temperature of 50,000 degrees being injected into the hot corona at 2 million degrees. X-ray images also show a profusion of spikey eruptions launching their way through the corona.

Quiescent prominences are altogether calmer affairs, not necessarily anything to do with the energy surplus in active regions. They occur in weakened, tired, old magnetic regions, and are long sheets of visible gas shimmering vertically above the solar surface. The bases of these luminous curtains are tied down to the borders of super-granulation zones. Unlike the energetic prominences, the quiescent ones have a leisurely life lasting a few months. Excellent photographs show that the structure of a quiescent prominence is mainly a series of fine vertical ropes (ropes that are 200 kilometres *thick*, incidentally), that gas slowly drifts through, the speed is 3,500 km an hour and that's slow for plasma in a prominence! Somehow new material is fed into the tops of these prominences, for they could not last so long without fresh supplies.

One of the greatest, certainly the most famous, prominences ever sighted took place on 4 June 1946. This one has certainly been unsurpassed in beauty. A giant arch spanning three-quarters of a million kilometres trembled in the chromosphere and corona and then suddenly began to ascend to higher altitudes before vanishing into the corona.

Prominences are usually photographed at the limb of the Sun, where they stand out as flame-like structures against a velvet black sky. But they can also be seen on the solar disc, where they show up as dark snake-like features, named filaments. Of course they are not completely dark, but they definitely are much less bright than the brilliant photosphere and hence they *seem* dark when viewed against the disc.

The spicules, encountered earlier in this book, may in many respects be considered as mini-surges or prominences. Shaped like cones, they have a diameter of 1,000 km (600 miles) and extend to ten times their diameter into the corona. On the Sun there are hundreds of thousands of spicules at any moment, each of which lasts for five or ten minutes before dissolving away.

A brief encounter with prominences can give the impression that they are hot flames flickering above the Sun. This is not really true, as we shall see. The prominences are in the lower corona, where the density of electrons is about 10^8 per cubic centimetre and the temperature is roughly 1 million degrees. Prominences, on the other hand, have electron densities at the very least 100 times larger—in the vicinity of 10^{10}–10^{12} per cubic centimetre, and temperatures of about 10,000 degrees. In other words prominences are much denser and much cooler than the corona; their electron temperature of 10,000 degrees inevitably means that most of their energy emerges in the optical spectrum. Not so in the corona, where the searing temperature has made the gas transparent to optical radiation and a potent source of X-rays.

Solar flares are another major phenomenon associated with the active regions; they are probably the most complex feature seen in the Sun's skin. When a flare takes place, astronomers see brilliant flashes of light in the solar atmosphere. They last for less than an hour, even for only a few seconds, as they flash away. Although the brightest flares can be seen when the Sun is imaged in white light, they are easier to detect and analyse if most of the normal sunlight is filtered away. It is usual therefore to look for flares through narrow-band filters that transmit only the spectral lines of hydrogen or calcium.

Hydrogen-light photographs will typically reveal a star-like brightening in the lower atmosphere, probably within a plage region. During a flare the solar atmosphere (mainly in the chromosphere) brightens across the entire electromagnetic spectrum. A sudden release of pent-up magnetic energy leads to temporary local heating of plasma. Electrons, protons and other charged particles get whipped into a frenzy as electromagnetic energy sparks off all around them. In next to no time the electrons are flying about at nearly the speed of light, and, by interacting with other charged particles, as well as the magnetic field, they are stimulated to emit energy of every kind—from very long wavelength radio waves right through to intensely energetic X-rays. The most impressive flares generally rampage through the regions where very large spots are festering.

Flares are the most important feature of solar activity to affect the Earth. The charged particles shot out during the flares crash against our planet's upper atmosphere. It is the flares that rip into the ionosphere, disrupting radio communications and creating auroras.

The ultraviolet and X-radiation from the Sun increases greatly when there are flares, for they are a high-temperature high-energy phenomenon. Our knowledge of the flares has advanced substantially thanks to the Skylab observations. A crucial aspect of that study was the continuity of images obtained during the mission. Astronomers were able to follow the history of flares back in time to discover that their birth takes place at the tops of tight magnetic arches looping out of the Sun. Measurements confirmed that the energy release at short wavelengths is indeed much greater than at optical wavelengths.

Ultraviolet spectra of the flaring regions showed as many as 5,000 different emission lines. On the first Skylab mission the scientists photographed a flare in the light of iron vapour at a temperature of 17 million degrees. It is probable that flare temperatures get up to 20 million degrees, hotter than the Sun's reactor core. This high level of excitement switches on the atoms in the chromosphere, causing the rich emission line spectrum.

Flaring rises and falls in time with the solar cycle as indicated by the spottiness of the disc. It springs into action, apparently, when excess magnetic energy has built up in an active region. This may happen because the magnetism has got twisted up, or pinched together over a sunspot pair. At some stage the tension wants to break, and this happens as field lines rapidly short-circuit each other and join up. As the lines reconnect, energy is released in a way that probably causes the tremendous accelerations observed, as plasma streams its way into the corona. According to data from space-telescope observations the actual flaring seems embedded in the top of the loopy structure above spot pairs.

The very speedy release of energy in a flare is puzzling to theorists. Somehow the magnetic field steadily acquires energy and, despite all kinds of little disturbances that must be there as gas thrashes around in the active region, it is able to prevent serious leakage of the excess energy. Then the energy so thriftily stored is dissipated in one great splurge in which as much as 10^{25} joules get released; this energy bank is the same as the energy emitted by the entire Sun in one-twentieth of a second, and equivalent to the total amount of solar energy striking planet Earth in an entire year! Ejection of mass during such an event can amount to 10 billion tons, propelled outwards at a velocity of 1,000 km per second, with individual particles getting up to half the speed of light.

Studies of the flares at optical observatories are complemented by the satellite and rocket measurements. As the flare bursts through the chromosphere and corona a shower of hard X-rays is triggered, taking less than a minute to reach peak intensity. The radiation is made by electrons that are first strongly accelerated and are then sharply decelerated as they collide with the gas outside the flare itself. Much of the X-radiation made in this way arises in the dense lower chromosphere. As the energy is dumped into the chromosphere the material already there gets heated explosively, like a bubble of gas being zapped by a pulsed laser. In no time at all, an expanding blast wave is hurtling away, faster than the escape velocity, taking with it 10 billion tons of the daytime star that will now be slung through interplanetary space.

Nothing is simple on the Sun, not even an explosion in the atmosphere. The blast wave takes some of the magnetic field with it, and these moving lines can function like giant atom-smashers or particle accelerators. Nuclear matter gets whirled up to very high energies, although, to be honest, no theorist really understands how it happens. One consequence of this is that protons and neutrons bash together hard enough to make deuterium; as I have mentioned, the gamma rays that are emitted when the shuddering deuterium particles settle down have been detected.

Another gamma-ray signal which has been detected from the Sun is caused by the destruction of pairs of electrons and positrons. The positron is the anti-particle of the electron, and when the two opposites get flung together—as happens apparently in solar flares—they snuff out in a flash of light. This flash is rather special for it consists of two gamma-ray photons with exactly the same energy, an energy characterizing uniquely the annihilation of electron-positron pairs. There are other gamma-ray lines detectable that probably result from collisions between charged nuclear material and neutrons.

Apart from the very energetic X-rays and gamma rays, softer X-rays with a thermal spectrum are also detectable. These are thought to arise in coronal gas in the neighbourhood of the flare that has been heated up to about 10 million degrees in the flaring process. A similar mechanism results in the ultraviolet emission seen to come from the chromosphere.

When a flare is in progress, solar radio astronomers have plenty to keep themselves occupied. The charged particles sloshing about in moving magnetic fields produce a good deal of noise in the radio spectrum. Large solar flares are frequently accompanied by outbursts of radio noise at metre wavelengths comprising what are known as type II bursts. (Different types of radio noise have been classified by radio astronomers as type I, II . . . bursts, but only types II and III seem to be directly associated with flares.) The emission becomes detectable at later times at lower frequencies, as a consequence of which special detectors had to be constructed. These radio bursts commence about ten minutes after the flash is seen and continue for about the same length of time. They are stimulated by streams of radiating particles accelerated in the flare. As the beam of particles ascends through the outer solar atmosphere it is able to excite radiation at progressively lower frequencies. In fact, spacecraft have detected these bursts right down at 300 kHz coming from a distance of over thirty solar radii beyond the Sun's disc.

The development of radio bursts with time has been beautifully mapped under the leadership of Paul Wild at the Culgoora Observatory in New South Wales. This fine Australian instrument is an interferometer of ninety-six elements. Its purpose is to map the radio waves from the vicinity of the Sun twice a second. The image of the radio Sun is made of a mosaic of discs, each of which is roughly two minutes of arc across, corresponding to a distance at the Sun of roughly 100,000 kilometres (60,000 miles). The radio telescope not only perceives the type II bursts as they expand into space but can also map the emission from giant prominences.

The Sun is, as we have seen in this chapter, a moderately variable magnetic star. We see the details of its magnetism to a degree that is unthinkable for other stars. At the same time, this magnetism is very feeble when compared to the truly magnetic stars: white dwarfs and neutron stars. All the same, this relatively weak magnetism has

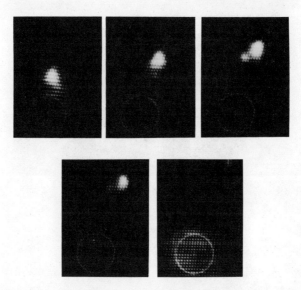

Five images from the Culgoora radioheliograph show the expulsion of a cloud of radio-emitting plasma from the Sun, the position of which is defined by the circle. (CSIRO Division of Radiophysics, New South Wales)

profound effects on the dynamical behaviour of the Sun's atmosphere, causing the rich variety of energetic phenomena: prominences, noise storms and flares. Beneath the photosphere the magnetism lies triggering the sunspot pairs. And as differential rotation—the engine that effectively winds up the field lines—proceeds, the Sun runs through its familiar cycle of changes. But has it always done so? The evidence is very strongly suggestive of a Sun that fluctuates on timescales of centuries.

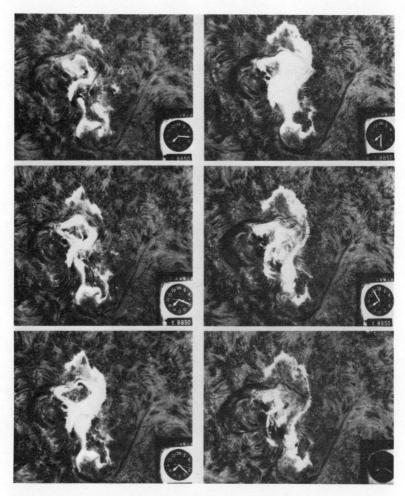

The great solar flare of 7 August 1972, photographed in the red light of hydrogen, is shown in a series of six images taken at intervals over a period of one hour. (Big Bear Solar Observatory, California)

Although we know more about the magnetism of the Sun than of any other star, we are still a very long way from seeing the how and the why of this magnetism. Although I have given a qualitative picture (glimpse would be a better word) of how sunspots and flares possibly work, most researchers would truthfully say that neither of these is understood well. We certainly do not have a full theory for the generation of the magnetic field in the first place or the mechanism for turning it over periodically. Many questions are still unanswered. How do flares store the magnetic energy before the bomb goes off? And what is the fuse for the bomb? The puzzles are at the forefront of research and thus they will continue to receive detailed consideration in the next few years.

On the positive side it can be said that the study of the active Sun has led to enormous advances in astronomy, astrophysics and plasma physics. In its early days the science of radio astronomy was able to develop not only scientifically but also in terms of its ability to get a larger slice of science budgets, due partly to the richness of the solar radio emission, a topic only touched on here. Plasma physics, the study of how hot gases move and behave in the presence of magnetic fields, has been strongly influenced by the findings in the solar atmosphere, which can be considered as a cheap accessible natural laboratory. Back in history the rediscovery of sunspots in the seventeenth century aided the essential change in the minds of men (make your brain respond to observations of nature and the dogma can take care of itself) that was to be the foundation of an entirely new scientific age. And we must not forget the infinitely fascinating radio galaxies and quasars, prolific sources of cosmic energy far out in the Universe. Perhaps just a little of what we have tried to learn so near to home can be applied to these exotic objects. Mechanisms of electron acceleration; containment by magnetic fields; the channelling of the energies of many individual particles into the enormous energy banks of a few particles, as in a solar flare—all these phenomena and more besides can probably find applications in radio galaxies, supernova remnants and X-ray stars.

One of the deep and searching questions we can ask is where the Sun's magnetism really comes from. The Sun is made of material with a high electrical conductivity, so the magnetism gets frozen into moving plasma. The fact that the churning gases carry magnetism with them makes the study of solar magnetism a hard problem area of astronomy. Various motions in the gas induce electric and magnetic fields, thus creating conditions suitable for setting up a dynamo inside the Sun. At the beginning of the twentieth century J. Larmor attempted a rudimentary theory of solar magnetism being sustained through dynamo action. W. M. Elsasser revived the theory in the 1940s, and it was further developed by E. N. Parker in the two following decades. Essentially the theory assumes that there *is* a field present and aims to explain how this field can be sustained, given that the material in the Sun is a good conductor of electricity. There is no serious problem in giving the Sun a bit of magnetism to start with. Magnetism pervades the Galaxy: when the solar nebula collapsed it could quite easily have taken a part of this universal field with it, compressing and amplifying it as the proto-Sun formed.

Dynamo theory seeks to show how conducting gas, sloshing around in a magnetic field, can generate electric currents that maintain the field despite its natural tendency to decay. This is a self-exciting or self-sustaining dynamo, and the possibility that it could work was established by E. C. Bullard in 1949. Parker's early attacks on the problem showed that certain kinds of magnetic field would be amplified by the swirling motions of the Sun's differential rotation. Later work linking the sunspot fields, differential rotation and the overall sunspot cycle, with a weak general field, has strengthened the faith of astronomers in the dynamo theory. Nevertheless it still does not rest on entirely firm ground, and will not do so until the mechanism is demonstrated beyond doubt by calculation and computation.

Into Space 10

The Sun loses some 4 million tonnes of its mass every second as hydrogen is processed through the central reactor. This vanishing act, the complete destruction of material, is not the Sun's only mass loss, for it also has to contend with the breeze of particles that rushes outwards as the solar wind. At its outer limits the corona is very hot but only weakly bound by gravitation; furthermore the rate of change of temperature is not fast. This leads to a situation in which the hot outer corona effectively expands steadily into the vacuum of space; this coronal outflow is called the solar wind.

The idea that the Sun might be throwing particles into space goes back to the 1930s at least, when two scientists, S. Chapman and V. Ferraro, developed a model for the collision of clouds of solar plasma with the Earth, in order to account for sudden changes in the magnetic field at the surface of the Earth. New clues came in the 1950s from careful investigations of comet tails. Astronomers had long known, of course, that as a comet swings around the Sun its magnificent tail always points away from the Sun. Theory had it that the sheer pressure of radiation streaming out of the Sun pushed the tails out from the comet head. However, comet spotters began to notice that the gas in comet tails sometimes made unaccountable sudden jumps as if something were crashing into the tail. In 1951–3 the German astronomer L. Biermann stated that these changes could be caused by particles streaming out of the Sun continuously. Today we know that Biermann's model of the interaction is incorrect but the basic idea of a solar wind flowing radially from the Sun is his. Finally, in 1958, E. N. Parker calculated that the hot corona simply could not sit around the Sun as a static shell; no way could it avoid expanding into space as a wind.

Direct measurements, confirming the existence of the wind, became possible in the space age. Russia's deep space probes, launched in 1959–61, detected charged particles flowing through space; American

Comet Bennett 1970. The tail of a comet always points away from the Sun, regardless of the direction of motion of the comet, as the solar wind and radiation force the cometary material away from the Sun. (Tautenberg Observatory, Tautenberg, DDR)

scientists confirmed this general result with the Explorer 10 craft in 1961. The remarkably successful probe to Venus, Mariner 2, removed any remaining doubt in 1962. For three months this sturdy craft ploughed onwards through the solar gale, recording average speeds of over 500 km a second—a million miles an hour.

The 1960s and 1970s were a remarkable period for solar-wind studies, for the sheer acquisition of data put astronomers in the position of knowing more about this breeze than any other plasma in the universe. But, as we shall see, the solar wind has also been examined by telescopes on Earth.

At various stages in the history of astronomy, the study of the Sun provided important information for non-solar astrophysics, and this is one of the continuing justifications for looking at the Sun in detail. In 1964 the reverse process occurred: the need to examine certain extragalactic radio sources by new techniques effectively led to the appreciation that the solar wind could help the research efforts on very distant galaxies. In the early 1960s much of the effort in radio astronomy was concerned with radio galaxies. These unusually disturbed galaxies are powerful sources of radio energy; they continue to mystify astronomers, even now, though not perhaps to such a degree as formerly. Some of the radio sources outside our Galaxy seemed to have very small angular diameters. A proportion of these radio sources lies close to the ecliptic—which can be considered as the apparent path of the Sun as viewed from Earth relative to the fixed stars—and these behave very strangely indeed when they are positioned on a sightline that passes close to the Sun. They exhibit the effect called scintillation. What happens, roughly speaking, is that the radio waves are being distorted as they pass through the plasma clouds near the Sun, just as objects viewed across the top of a hot surface, such as an electric cooker heating element, appear to shimmer and shake through the rising bubble of hot air.

Further investigation has revealed several interesting properties of the solar wind, and these studies have supplemented the direct probing with the IMP satellites. The solar wind helped astronomers in a truly remarkable way: the telescope built at Cambridge to study the scintillation of radio galaxies and quasars, caused by the solar wind, led to the discovery of the flashing pulsars, first detected in 1967.

The dynamics of the solar wind are intimately linked to the corona and its magnetic field. This is because the wind has a very high electrical conductivity, which makes it more or less impossible for it to flow across magnetic field lines. Much solar magnetism is actually dragged along as the breeze flows away. There are large variations in the expansion velocity, which is evidently strongly influenced by solar activity and hence the amount of energy being fed into the corona as heat. As measured at the Earth the flow can rise from low speed of 400 km (250 miles) per second—to a higher speed of double that value in a couple of days. Those streams, as they are called, then typically relax back to the lower speed more gradually, having maintained the peak flow for several days. As is usual in astronomy, the velocities are quoted as distance per second; let's not forget that 400 km per second is over 1.25 million km an hour—some gale! When a new stream is racing out from the Sun it is detectable by a spacecraft as a sudden increase in the magnetic field and particle density outside the craft. Then a few days later these quantities both sink to unusually low values, just like the peaks and troughs of a wave motion.

As we saw in the last chapter, when a great flare takes place on the Sun, extra material is ejected from the corona at high speed. As this stuff moves rapidly outwards it ploughs into the slower traffic that is observing the standard solar-wind speed limits. This causes a pile-up, termed a shock wave, that is several million kilometres thick. Eruptive prominences are another source of extra material that is poured into the general wind flow.

The fact that the solar wind drags magnetic field with it has interesting consequences. For one, there is an interplanetary magnetic field which is maintained by the wind. Within about three solar radii of the Sun's surface the magnetism is strong enough to beat the wind energy; in other words, the flow of the wind is governed by the local magnetic field. Beyond this critical radius, however, the wind is in charge. As a result the fields are contorted and get blown out through the solar system. As far as the critical radius, the corona has to rotate along with the rest of the Sun, but beyond it slippage takes place, which you can picture as happening because the magnetism is no longer strong enough to grip the outer corona in place as the Sun rotates beneath it. The end result is that the coronal magnetic field lines get strewn out in a spiral structure, rather as happens with water when your garden hose goes out of control and whirls madly round; the same effect is given less dramatically by certain types of rotating lawn-sprinkler.

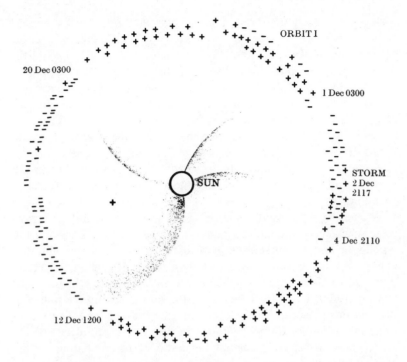

The interplanetary magnetic field is divided into sectors of opposing magnetic polarity.

One intriguing property of the wind and the magnetic field is that the flow is not strictly radial (that is, not like the spokes of a bicycle wheel). Viewed from one of the poles of the Sun, it is swirling out at a slight angle. This means that the wind is carrying with it angular momentum from the Sun. The solar wind is acting as a brake which is relentlessly decelerating the rotation of our daytime star. This is most interesting because one of the mysteries of the Sun is why it rotates slowly at all. After all, the Sun condensed from part of an interstellar cloud, which, although it might seem to be wafting gently through the Galaxy, would have been twirling rapidly once it had condensed to a proto-Sun. In some way the Sun has rid itself of most of the rotation energy it once had. One plausible way it could have done this is by a slingshot effect as it hurled the solar wind along its magnetic tentacles.

A simple calculation shows that the solar-wind brake slows the Sun right down in about 5 billion years. That's the Sun's present age! So the spin rate today is substantially less than it was at the Sun's birth. There is also the possibility that magnetic links between the proto-Sun and the proto-planets enabled the solar system to arrange for almost all the total angular momentum of the entire system (98 per cent) to be vested in the planets rather than the parent star. This may explain why the Sun has less spin energy than many stars; it transferred much of its rotation energy to the planets very early on. However, the true pattern of events in the solar nebula is now lost in the mists of time, but we can say that the wind continues to act as a gentle brake.

What's the solar wind made of? The Apollo lunar missions gave solar scientists the chance to measure its chemical make-up. On the Apollo 11 and 12 missions a piece of aluminium foil was unfurled and placed like a banner facing the Sun. While it was on the Moon each foil received a constant bombardment of particles in the solar wind. Before heading for home the astronauts rolled up the foils and brought them back to Earth. Laboratory analysis of the foils gave an accurate reading of the atoms found in the solar wind, a remarkable achievement because it amounts to a direct determination of the composition of Sun matter. Scientists deduced that the number of helium atoms relative to hydrogen atoms was roughly one in twenty. In terms of mass this means that about 15 per cent of the mass in the solar wind is helium and almost all the rest is hydrogen.

Now we have to relate this quantity to what theorists think they know about the Sun's insides. The cosmic helium abundance is generally taken as about one nucleus in every ten being a helium nucleus; the corresponding mass ratio is 25–30 per cent helium, which is twice as high as the value for the solar wind. Helium seems to be less common in the outer corona and solar wind, or we could conclude instead that the measurements are wrong. Probably what happens is that the Sun is able to hold on to its atmospheric helium better than the hydrogen, so we get the misleading impression that helium is deficient. Other substances identified in the solar wind are oxygen, carbon, neon, silicon and iron, detected by Vela spacecraft in the late-1960s.

Within the general flow of the solar wind there are variations, particularly when high-speed streams get switched on. These were first picked up in 1962 by the US spacecraft Mariner 2. When this probe encountered a stream, the average flow speed doubled from 300 km per second to nearly 600 km per second in about two days; this then declined over a period of five days. Observation carried out by subsequent spacecraft established that these fast lanes for the particles wishing to leave the Sun, rotate with the Sun. This ties in with an important result from the IMP 1 interplanetary spacecraft in 1963. Measurements of the direction of the interplanetary magnetic field beautifully demonstrated that it is broken into sectors that rotate with the Sun. Inside the large sectors the magnetism has a clearly-defined polarity, and this is maintained at successive solar rotations. The sector pattern would imply that the solar wind must, in its turn, arise in sectors of the corona which are similarly organized magnetically. One neat result from IMP 1 was a matching of the magnetism in interplanetary space in a way that allowed it to be compared with the magnetism of the photosphere. After allowing for the time-lag, while the solar wind swept the field along from the Sun to IMP 1, scientists achieved a beautiful match between the disc field and the interplanetary field. This reinforced the view that solar magnetism, the solar wind and interplanetary magnetism are interwoven.

The high-speed part of the wind is formed into tubes that spiral out through the solar system. At the sunward end these are apparently anchored to points on the Sun where the magnetic field is streaming radially outwards. But these do not seem to be closely related to the solar active regions. Intuitively we might expect the sunspot and active areas to be the engine-driving jets in the solar wind. Not so: there is no clear link between common solar-wind phenomena and changes in the numbers of active regions.

Skylab established that the Sun's polar regions are an important source for the fast streams observed in the solar wind. Observations made during and immediately after the missions established that the high-speed wind flows away from the polar caps. In these two locations the magnetic field lines open away from the surface, allowing plasma to flow away freely. The shape of the magnetic field near the Sun is such as to channel some of the fast streams down into the main plane of the solar system, which is where the spacecraft measurements are made.

Solar flares have a very definite effect on the wind gusting through our planetary system. Imagine what happens: the magnetic short-circuit that triggers a flare results in an enormous amount of energy discharging into the corona. The debris from the flare—high-speed electrons and protons—crashes through the general wind and interplanetary fields, creating a shock wave where it is compressing the local gas. Spacecraft, such as those in the Vela series, are able to monitor these shock waves because many of the instruments on board will record abrupt changes in speed, density and particle temperature as the wind whisks past the craft.

In addition to the wind of atomic particles, interplanetary space also contains solid specks of dust, as well as the gaseous stuff that makes up the solid wind. This dust is responsible for a beautiful Sun-related phenomenon that is best seen in the clear skies of moonless nights in the tropics. The first time I saw a really transparent sky far away from city lights was in New South Wales, several hundred kilometres south of the tropic of Capricorn. Here the luminous character of the sky, compared to that in England, astonished me; standing inside the dark dome of the Anglo-Australian Telescope, the small patch of sky visible through the dome aperture seemed absurdly bright—the night sky is, after all, supposed to be dark! The striking feature responsible for the bright night sky is the zodiacal light, sunlight being scattered from dust in interplanetary space.

J. D. Cassini commenced a ten-year study of the zodiacal light in 1683, and correctly deduced that it arises from the reflection of sunlight off dust particles. These are collected into a thick lens-shaped cloud that is centred on the Sun and is symmetrical about the main plane of the solar system. Further thoughts, published by D. de Mairan in 1733, put the correct view that the cloud of spacedust extends at least as far as the Earth's orbit. Under superb conditions the zodiacal light reaches in a cone up to 60° from the Sun (which is, of course, below the local horizon). Remarkable as it may seem, our knowledge of the 'false dawn' of poetry and literature is practically the same three centuries after Cassini: we may have more data, but his model is essentially right.

Astronomers study the zodiacal light for several reasons. One is that it is the brightest *extended* source of light in the sky at low terrestrial latitudes, which makes it easy to measure. And, incidentally, it's by taking the spectrum of the light, which shows the Fraunhofer lines, that we can be sure it really is scattered sunlight and not radiation from an unrelated source. Another reason is that it gives us a method of finding out more about spacedust without having to send a probe out there. Since spacedust is cold and dark, the only source of information on its large-scale properties comes from analysing the scattered sunlight. Most of the particles are, in fact, something between 10 and 100 microns in size; a micron is a millionth of a metre. The dust is like extremely fine powder, much finer than sand, and it would cause choking if you breathed it in quantity.

Spacecraft gliding out to Jupiter have looked at the zodiacal light properties beyond the Earth's orbit. Right through the asteroid belt the light is still detectable, but when the craft has reached three times the Earth's distance from the sun, no further light is detectable. The mere fact that the dust, and the dim light bouncing off it, extend well beyond the Sun implies that it should be possible to trace the back-scattered sunlight right across the dark night sky. This is indeed the case. If you scan a photometer along the ecliptic the signal steadily falls at larger angular distances from the Sun. But at about 150° from the Sun the signal rises instead and continues to do so until the photometer is aimed at the point in the sky exactly opposite the Sun.

At large angular distances from the Sun we encounter the so-called *gegenschein* (literally this word means counterglow), a very faint glow of light opposite the Sun, caused by back-scatter from the dust. I have never seen the gegenschein, which is much fainter than the zodiacal light. People who have seen it say it is 5 to 10 degrees across, oval in shape, with its longer axis along the ecliptic.

Space is dusty mainly because of the comets that come to the inner solar system. As we have noted, encounters with the Sun drive off gas and dust, which stream out behind the comet as it unfurls in the solar wind. Comets, whatever they are, are almost certainly extremely primitive solar-system members, dating from the early condensation of solid objects orbiting the half-formed Sun. Many questions about comets could be settled by launching a spacecraft to intercept one; this expensive technique must be limited to those comets whose arrival can be predicted well in advance, on account of the long planning time needed for space missions, and, at the time of writing, it seems unlikely that we shall try to intercept Halley's Comet in 1986. Without doubt the comets contain a great deal of dust, and, as shown by investigations of their tails, they sprinkle this liberally about space as they voyage through the solar system, leaving the greatest trails when they are closest to the Sun.

Studies of the Sun's outer environment in the 1970s were particularly fruitful as a result of the Skylab missions. Already, the OSO-7 craft has revealed the great complexity of the corona when imaged in high-energy radiations. The X-ray photographs showed a scattering of bright points and dark holes in the corona, together with high-reaching arches and loops of magnetism. The structure of the outer corona is, indeed, dominated by magnetism, which sculpts the fingers of plasma streaming from the Sun, and moulds them into plumes, helmets and flame-like protuberances.

As Secchi probably realized, in 1875, the structure of the corona varies with the solar cycle. By 1896 C. A. Young had tied down the distinguishing features of the maximum and minimum types of coronal architecture. During the minimum phase the corona is not conspicuous, as we have already noted. The relatively faint light is given structures by streamers of plasma that emerge in the low latitudes (near the solar equator) and short plumes. The streamers are located over plage regions.

During periods of solar activity we see that over the poles of the Sun there are polar plumes during the intervals of minimum solar activity. These vertical columns of plasma are as much as 8,000 km (5,000 miles) across and they extend to the amazing altitude of half a million kilometres. Structures named helmets are also seen over plage areas; their upper ends draw to a sharp point at one or two solar radii above the surface. Huge active streamers will sometimes grow over sunspot regions; these can be traced on photographs taken at eclipse out to five solar radii and more. These superb modules of coronal structure—helmets, plumes, arches and streamers—are all effectively shaped by the outer magnetic field.

The fantastic richness of coronal structure is only brought home to us in the ultraviolet and X-ray images, such as those obtained by Skylab. As we have seen, the corona is extremely hot, and effectively transparent to optical radiation; in eclipses we get a sideways view. To get the face-on picture you must go to X-rays, for the stripped-down atoms in the corona mainly emit ultraviolet and X-radiation. These atoms, more properly called ions, have lost their outer shells of electrons in the searing thermal environment. Their remaining electrons make large jumps between energy levels, jumps that lead to the emission of giant-size energy packets, or X-ray photons. Another vitally important factor is that the nature of the X-radiation is sensitively dependent on both temperature and electron density; by turning this statement on its head we can see that mapping the corona in a range of X-ray energies will allow its temperature and density structure to be deduced. In fact, the X-ray intensity is governed by the *square* of the electron density, whereas white-light intensity depends only linearly on the same quantity. This is one reason why the X-ray images show hot or dense regions in the corona with very good contrast.

In an X-ray photograph of the solar corona large and bright areas overlie the active photospheric regions. Evidently the strong and complex magnetic field is also determining the energy flow into the corona above the active region. The loops connecting regions of opposite magnetic polarity show up extremely well.

Small bright spots of X-ray emission are seen. A comparison of an X-ray photograph and a magnetograph obtained at the same time readily shows that these X-ray pinpricks are associated with magnetic bipolar regions. The magnetic regions that cause the X-ray bright points are so compact that the tight loops of magnetism that channel and imprison the hot X-ray gas cannot, with current instrumentation, be distinguished from actual point sources. The X-ray sparks don't last all that long, fizzling out within a few hours. The bright points must be related to conventional active regions, but for some reason they live for a much shorter time. Just as with solar flares they switch on rapidly. But there is one very important respect in which they differ from active regions; that is, they are sprinkled all over the Sun, not confined like sunspots to the tropical zone of activity. Astrophysicists speculate that the X-ray bright points are significant contributors to the magnetic flux dragged out of the Sun. They probably tug out as much magnetism as the conventional active regions. There are so many bright points, scattered like jewels across the Sun, that they may even be the dominant harbours of solar magnetism.

The most interesting discovery made by high-energy studies of the nearest star is probably the finding of holes in the corona. Coronal holes are the regions that look black in the false-colour or monochrome X-ray photographs; they are large volumes of the corona that choose not to emit any X-radiation. Why should this be?

Coronal holes were pursued in the early 1970s, initially by means of ultraviolet observations. These pioneering rocket studies showed that their designation as holes is apposite, because the density of hot gas in a coronal hole is only one-third the normal value for the quiet Sun. Nor is that all: the temperature is about one-half that indicated in the rest of the corona. The zone that marks the transition between the chromosphere and corona is much thicker beneath a hole.

In the days when astronomers could only stand on the Earth, the holes were well hidden. Beneath an atmosphere that absorbs high-energy radiation, and staring at the light of the daytime star, astronomers could gain no clue to their existence, for holes have hardly any effect on the photosphere or the lower chromosphere. All the frothy surface, its granulation and supergranulation, the shivers and shakes of the oscillating Sun, these are no different beneath a hole. The one valuable clue comes from considering the magnetic field, which opens out to the solar system inside coronal holes. In essence a coronal hole is a very large area of the corona that is cool and of low density. Within it the weak magnetic field spreads out from the Sun. The holes are therefore an important source of the solar wind.

Let us now look at the energy balance in a coronal hole in some detail. The corona generally is the very rarefied, very hot, final layer in the solar onion. It transfers radiation to the remarkably cold Universe, only three degrees above absolute zero, by radiation. The heat energy needed to maintain the coronal temperature up at the two-million mark comes from mechanical waves, the bangs and buffetings, blasted at the corona by the chromosphere. Various phenomena associated with solar activity add to that energy input. In a stable situation, and the corona is effectively stable, what goes in must flow out. Now we can perceive a problem: holes are colder, nearly a million degrees colder, than the rest of the corona, so they cannot radiate nearly so effectively. Furthermore, the smaller rate of change of temperature through the transition zone means that heat is not conducting from the corona back down to the chromosphere at anything like the normal rate. We see that a coronal hole apparently has a plug at each end. One of these slows down heat flowing by conduction from corona to chromosphere, and the other eases off the rate at which it flows from corona to frigid Universe. Yet holes are stable (Skylab saw them endure for nine months), so they must be ridding themselves of energy somehow.

The way out of this problem is provided by considering further the effect of the open, or diverging, magnetic field on the solar wind. In the open regions, such as the holes, the wind flows out easily because it does not have to take any web of magnetic field with it. The excess energy is not bottled up at all: the Sun uses it to push the solar wind out through the coronal holes, the major source of the wind. There is a precise connection between the passage of coronal holes across the disc of the X-ray Sun, and the arrival on the Earth of the high-speed particle streams that are present when the wind is blowing hard. Skylab scientists, observing the coronal holes, established that when you make allowance for the several days that it takes for a gust of particles in the wind to reach Earth, the match between holes and particle streams is perfect. Hence the holes definitely have a profound influence on the wind, and they also indirectly cause changes to the Earth's own magnetic character, and hence are of practical importance to us on Earth.

The picture from above the poles of the Sun may be more interesting still. At present we cannot easily peek over the north or south pole of the Sun, although it is possible to work out a little of what is happening by straightening out geometrically the drastically foreshortened view that we terrestrials get of the Sun's poles. It happens that there are holes over the poles too. During the Skylab programme of 1973 one of these holes could be observed for eight months. This was so persistent that it must have been a very effective source of solar wind. Radio astronomers making observations of distant galaxies and quasars confirm that there is a wind gushing from the poles. It's quite possible that this main polar flow is a good deal more impressive than the fraction which is channelled into equatorial flow that we and our spaceprobes, in the plane of the solar system, are limited to measuring. There is an intriguing proposal to send a spacecraft up and over the Sun to look at one of the poles. This out-of-the-ecliptic probe would probably be aimed first at the giant planet Jupiter. Then, in a game of interplanetary billiards, Jupiter's strong magnetic field would fling the craft around in a close encounter, and swing it up from the ecliptic plane. Then it will be possible for astronomers to make direct measurements of the density and velocity of the wind over the poles.

Sun and Earth 11

Obviously, the Sun is the one heavenly body which has the most influence on the Earth as perceived by mankind. In this chapter we shall look at some of the ways in which the radiation and particles from the Sun affect the Earth, its atmosphere, and even ourselves.

The heat and light from the Sun warm and illuminate spaceship Earth, which would simply be a frigid rock coated in ice if it were moved out through the solar system to ten times its present distance. Beyond these familiar warming effects the Sun also influences Earth in more subtle ways: it modifies the outer layers of the atmosphere, it shapes the magnetic field in the vicinity of the Earth, and it creates the stunning visual effects of the aurora. We call the branch of astronomy that tries to understand the complex and manifold interaction between Earth and Sun *solar-terrestrial physics*.

About 80 kilometres above the continents and oceans there begins a layer of our atmosphere named the ionosphere, which may extend out to 1,000 kilometres. In this region the energetic radiation from the Sun, as well as the natural cosmic radiation (high-energy particles from far beyond the solar system), collides with the atmospheric atoms and molecules. Ultraviolet and X-radiation, as well as the high-energy particles, have enough energy to prise electrons away from the atmospheric atoms or molecules and set them free. This part of the atmosphere is, therefore, *ionized*; it is made of electrically charged atoms and molecules together with free electrons. The region is oxygen-rich and has a high temperature—over 1,000 degrees. But the air here is so thin that despite this high temperature it would not cook anything; the temperature is to be regarded as a measure of the speed with which the ions and electrons are zipping around. As the Sun is the primary source of ionizing radiations, various measurable properties of the ionosphere change with the degree of activity on the Sun. If the Sun wants some peace and quiet then both the electron density and the extent of the ionosphere are reduced. However, a major solar flare changes all that, and gets the ionosphere boiling with activity.

Before the rocket age the ionosphere could only be explored by radio waves. In the present era of satellite communication for the relaying of telephone, radio and TV circuits between continents, however, it is easy to forget that long-distance radio broadcasts once relied entirely on the ionosphere. Because this region has abundant free electrons it is a good conductor of electricity, and in consequence of this long-wavelength radio waves get reflected from it, just as they would from a metallic screen. Over-the-horizon radio communication is thus accomplished by bouncing radio waves from the underside of the electrically-conducting ionosphere. Such communication is rather troublesome because the ionosphere layer varies with time of day, the seasons and solar activity.

It is mainly in the ionosphere that the cosmic radiations (particle, X-ray and ultraviolet) that harm human life are filtered out. Sometimes we say that the ionosphere protects us from the damaging effects of solar radiation. Although the upper atmosphere does afford this protection, it is also important to realize that the complex life on Earth today has evolved from much simpler forms in an environment that has little ultraviolet or X-radiation. If the ionosphere had afforded less protection, then presumably life would have evolved differently so that organisms would have had more protection built in at the design stage. We indeed have simple examples: dark-skinned races developed in the tropics so the body could filter out the ultraviolet rays not already stopped by the air. Fair-skinned people can manufacture the dark pigmentation if they expose the bare skin to strong sunlight—and if you're fair-skinned you've probably discovered in a painful way that the build-up of the dark pigment takes a few days! Since life has evolved under this protective blanket, we have no natural defence against the naked Sun. For this reason, as well as others, space travellers need special protective shielding in their craft and clothing. The dosage of radiation received by air crews on very high-altitude supersonic planes should be continuously monitored by airline medical staff. Passengers are at less risk, even in times of great solar activity, as they make many fewer high-altitude journeys.

A major solar flare greatly increases the harmful particle radiation in the vicinity of the Earth. The highest energy particles are protons

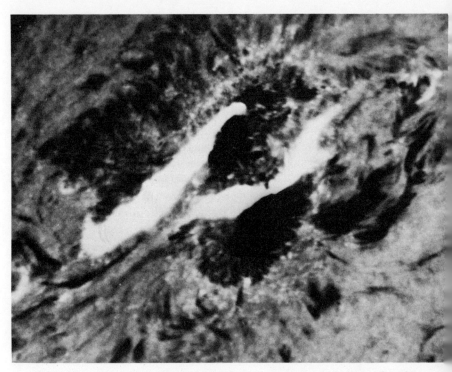

This solar flare of 10 September 1974 is a typical example of the phenomenon. The white light shows the flare itself, which is a potent source of high-energy particles from the Sun. (Sacramento Peak Observatory, Sunspot, New Mexico)

that get shot out of the Sun at just about the speed of light, as part of the active region acts like a particle accelerator or atom-smashing machine. These relativistic protons hit the Earth almost at the instant the flare is spotted by our telescopes. They smash into the atmosphere and collide with great force with the atoms. Showers of neutrons get produced and these are detected by instruments on the ground. A good solar flare causes an increase of ten or twenty times in the neutron count at ground level. These neutrons won't hurt you, but the speed-of-light protons would not take long to kill you off, painfully. For this reason it is usual to keep a careful watch on the Sun's temper when astronauts are required to work in space or on the Moon while protected only by a spacesuit.

The Sun and its variable radiations are responsible for some of the phenomena that plague amateur radio hams. To give just one example, short-wave fade out. This is a sudden cut off in the propagation of short-wave broadcasts. It happens when the Sun has caused increased ionization in the lowest layer of the ionosphere, which absorbs the signal. At very low frequencies the reflecting properties are increased, so that the low-frequency electromagnetic waves generated in natural thunderstorms travel large distances easily. This leads to a great increase in the number of thunderstorms that crackle away as atmospherics in a radio receiver.

Above the atmosphere and ionosphere, the effects of the Sun are extremely important in that intangible magnetic cage, the Earth's magnetosphere, which forms our planet's armour-plating against the ceaseless bombardment of atomic particles in the solar wind. The magnetosphere is the result of an interaction between the Earth's intrinsic, or personal, magnetism, and the magnetism and electric currents borne by the solar wind.

Let's begin by looking at the magnetic Earth. At present the field detected at the surface of the Earth can be reasonably modelled by assuming that our planet has a dipole (bar magnet) more or less at the centre. It doesn't actually have a permanent iron bar-magnet inside: rather the field is probably generated and maintained by the flow of electric currents within the Earth's fluid core, but the observed effect at the surface and beyond is similar to the field of a bar magnet. The Earth's dipole is tilted at about 11° to the rotation axis and misses the exact centre of the Earth by around 500 kilometres. The result is a magnetic north pole that emerges in Greenland and a south pole in Antarctica. In modern times the strength of the field has dwindled at a fairly slow but steady rate. If this continues the field will be zero in about 2,000 years' time. We know from studying the fossil magnetism trapped in rocks that the strength and direction of the geomagnetic field have varied throughout geological time. The Sun's magnetism switches every eleven years as its dynamo gets reorganized. Within the Earth, changes in the magnetic properties take far longer and do not seem to be regular.

The magnetic field measured at any point on the Earth's surface is composed of the intrinsic field plus the magnetic field, associated with the Sun and material from the Sun, that is crashing into or brushing past the Earth. Since the stormy Sun can change in a matter of minutes the measured geomagnetic field is not absolutely steady in either strength or direction.

The horizontal component of the field changes dramatically when a large decrease of its strength takes place. These sudden collapses, known rather picturesquely as *geomagnetic storms*, may last for a few days. During this time a sensitive magnetic compass might seem to be behaving somewhat erratically. We now know that these storms are not caused by sudden changes within our own planet. Instead, the magnetic Sun is the culprit, for the fluctuations in the field (and the compass needle) are caused by the arrival at the Earth of high-speed clouds of solar plasma, generated in a major solar storm. Active regions may persist on the Sun for more than one solar rotation. In that case the associated geomagnetic storm may also recur after about twenty-seven days of solar rotation relative to the Earth. Major geomagnetic storms are also associated with disturbances in the ionosphere that cause disruption in radio and TV transmissions, since both are caused by essentially the same solar phenomena.

As the solar wind brushes steadily past our planet it creates a region that moulds and confines the geomagnetic field which would otherwise extend far into space. The magnetosphere is large relative to the Earth itself. On the side facing the Sun it has its boundary about ten Earth radii beyond the Earth. There is an outer boundary layer called the magnetopause which is around 100–200 km thick. On the night-side of our planet the magnetosphere really does stretch a long way—1,000 Earth radii—like the tail of a comet. In fact it merges imperceptibly with the interplanetary magnetic field. At this stage it may be useful to fix in your mind the fact that the magneto-sphere is the magnetic skeleton around the Earth. Immediately be-yond this there is yet another special interaction region called the magnetosheath—the flesh on the skeleton.

The magnetosheath is the zone where solar wind particles are breezing past the magnetosphere, which they are scarcely able to penetrate. At the front end of the magnetosheath and facing the Sun there is the standing bow shock. This is analogous to the shock wave or sonic boom that accompanies a supersonic airliner. In the natural case of a magnetized planet sitting in the solar wind a shock front develops because the wind is flowing past the planet faster than the sound speed within the wind. This is just the same physical situation as the shock wave when an airliner moves faster than the sound speed in air. One special feature about the magnetic shock wave near the Earth is that it is of a type which is very difficult to create in the laboratory: the *collisionless hydromagnetic shock wave*.

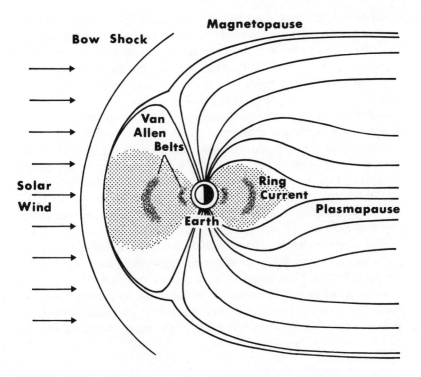

The Earth's magnetosphere is the terrestrial magnetic field deformed by the solar wind. Effectively our planet is surrounded by a cage that magnetically deflects charged particles from the Sun.

The deep space missions Pioneer 10 and 11 and the Voyager craft all visited Jupiter. This planet has a magnetosphere that is far more extensive than the Earth's, and has provided independent information on the nature of planetary magnetospheres. Saturn also has a magnetic field extending into space.

The portion of the magnetosphere that faces away from the Sun is variously called the *geomagnetic tail*, the *magnetotail*, or more plainly, the *magnetic tail*. The tail is rather like two tubes pressed together. In the upper tube the magnetic field points to the Sun, and in the lower one away from the Sun. When the two tubes meet there is a zone of neutral electrical charge because the oppositely directed fields cancel each other, more or less.

Of course, the magnetosphere is not a totally impenetrable barrier —particles do stray near to the Earth and we have already mentioned the effects they have on the ionosphere. The motion of charged electrical particles in the bar-magnet field of the Earth is such that particles with an appropriate energy may get trapped and orbit almost endlessly in the Van Allen radiation belts. The inner radiation belt was discovered and its shape deduced by J. A. Van Allen in 1958. Simple charged particle detectors aboard the first American artificial satellite, Explorer 1, failed to register above a height of 1,000 kilometres. Later, laboratory tests and further satellite observations showed that the null reading was in fact a result of the detectors' being totally overloaded in the radiation belts. The inner belt is mainly occupied by protons, whereas a more extensive outer belt imprisons electrons as well.

The trapping in the radiation belts occurs because the charged particles are made to spiral along magnetic field lines by electromagnetic forces. Near to the magnetic poles the lines are funnelling together and this puts the squeeze on the spiralling particles. A magnetic mirror is the result: particles get flicked back and forth from pole to pole, taking just a few seconds (at most) for each trip. In

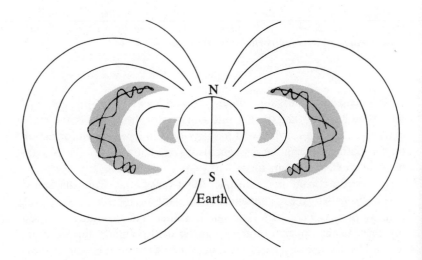

The aurora, a beautiful reminder of the constant interaction between the Sun and the Earth. (L. Snyder, University of Alaska)

order to understand more clearly the nature of these belts, experiments that now seem irresponsible were carried out in 1958 and 1962. Nuclear bombs were detonated in space in order to make several artificial belts of charged particles, but eventually such tinkering in outer space was stopped by international agreement. The Starfish explosion of July 1962 produced a radiation belt that persisted for several years. The folly of this exercise was brought home forcefully when it was realized that several expensive satellites had thereby been effectively wrecked because their solar power cells were damaged.

Another aspect of the physical relationship between the Sun and Earth is expressed by the beautiful shimmering lights of the aurora. The auroral displays have in fact been linked with magnetic activity on Earth since the eighteenth century. Now we are aware that the active Sun is responsible for both, and that the sunspot famine of the Maunder minimum was accompanied by a lack of auroral sightings from 1645 to 1715.

Most of the light in auroral displays is produced by atoms of oxygen and molecules of nitrogen that get excited in close encounters with low-energy electrons. Disturbances in the magnetism of the magnetotail shake electrons out of the tail and towards the Earth, where they get channelled down at high latitudes towards the magnetic poles. The electrons are actually compressed into thin sheets, and this gives rise to the characteristic drapery form of many displays. In fact it is hard to classify the actual forms of aurora, although researchers refer variously to arcs, bands, rays and veils. The size can vary enormously. Usually they occur at heights of 100–150 km (about 65–100 miles) and the horizontal extent can be anything from tens of metres (auroral rays) up to thousands of kilometres (arcs or bands).

The auroral zones, in which maximum activity is seen by observers on the Earth, are located in latitudes 67° north and south, and they are about 6° wide. The actual extent of the oval region around the magnetic pole in which the displays take place varies. At night it is usually 22° or so away from the pole; this contrasts with the general, but incorrect, view that the aurora takes place over the geomagnetic pole. Rather it is in a wide oval encircling the pole.

Because of the link between the auroral displays, the properties of the magnetosphere, and solar activity, the displays are governed by the sunspot cycle, the 27-day average rotation cycle, the time of year, and the general level of magnetic unrest. As a generalization, spectacular displays occur close to the maximum in sunspot activity. I once saw an amazing auroral display of shimmering green curtains, at moonlight, in latitude 67° north, while flying from Los Angeles to London via the polar route which crosses over southern Greenland.

The colour of the aurora is normally red or green. The red light arises from oxygen atoms and the green comes from nitrogen molecules. Emission is also detectable in the ultraviolet and infrared.

The Sun, then, has noticeable effects on the Earth's magnetic blanket. But what of the atmospheric cocoon? Surely the Sun must have effects on this too. Recently there has been an impressive upsurge of interest in the changing climate of the Earth, the causes of climatic change, and the prediction of future climatic trends. Although it is generally agreed that astronomical influences *may* have a major effect on the climate, there is no definite proof that any single climatic 'event' in the past, such as an ice epoch, can be attributed to an astronomical cause. One problem, of course, is that we only have good data on solar phenomena for the last three centuries, and careful measurements of the solar luminosity span over less than a century. These time bases are too small compared to the response time of the Earth's climate factory, which seems to produce climatic behaviour characterized by longer (several centuries) trends. Therefore, in order to gain a long enough span of data we have to rely on pre-instrumental information about both the climate and the astronomical phenomena. It has seemed to me, as an astronomer and not a climatologist, that getting sufficient reliable information on the past performance of both the climate and the Sun is hard, although I do not in any way want to imply that the improvement in our knowledge has been other than impressive. There is no doubt that the climate has changed and does change. Geologists have identified several major ice epochs in the last 3 billion years, during which great ice sheets covered much of the continental land masses. In geological terms the ice has only recently retreated. But the cause of the ice epochs is still a matter of considerable debate, and we do not know whether astronomical or solar influences have any bearing on the problem.

Since this is a book on astronomy I will first review the evidence for a changing Sun before seeking a link between Sun and climate. We have already met some of this evidence in earlier chapters, when discussing the solar cycle. The evidence for an inactive Sun during the seventeenth century, the period of the Maunder minimum, is very strong and based on historical records. In the 1960s a new technique became available for plotting solar variation back for a few thousand years. This is based on measuring the amount of radioactive carbon in old trees.

Radioactive carbon, or carbon-14, is made at the top of the Earth's atmosphere where high-energy charged particles from remote space smash into the atmosphere. When the Sun is active and spotty it has an extended magnetic field. This shields the inner solar system from the high-energy cosmic rays. When the Sun is quiet its magnetism affords less protection. More high-energy particles get through to the planets and more carbon-14 is present at the top of the atmosphere. This form of carbon has a half life of 5,730 years, so if it is isolated and stored up somewhere (such as in a tree) scientists can still work out how much was there initially, provided it hasn't been locked away for more than a few thousand years.

Trees do this storing for us. The growth of living plants requires them to absorb carbon dioxide from the atmosphere and, under the influence of sunlight, manufacture cellulose (an organic molecule containing carbon) as new-growth wood. This cellulose laid down each year in tree rings has a little carbon-14 trapped alongside the normal carbon. Long-lived trees, such as the famous bristlecone pines of California, may thus keep a diary for several thousand years.

There are limitations on this method. Eventually the record is washed out by the normal decay of the carbon-14, so that in practice it is only possible to glean data for the last 7,500 years from tree rings. A more serious problem is that the carbon-14 is made at the top of the atmosphere whereas the trees are growing at the bottom. Circulation of the radiocarbon is therefore very complex and, in particular, absorption and dilution in the oceans smear out variations on timescales shorter than twenty years. As a result of this there is no clear indication that the solar cycle of eleven years can be traced in the radiocarbon data. A major triumph of the method, however, was that the radiocarbon measurements showed the Maunder minimum clearly as well as an earlier suspected period of inactivity in AD 1450 –1540 and a period of marked activity in the twelfth century. Major features lasting for several solar cycles, and within the grasp of uncertain historical records, seem to be retained by the radioactive diary in the trees, so it may be reasonable that we should try to read the diary beyond the point where there is written back-up evidence.

One feature that immediately shows up over a 7,500-year time span is the effect of the slowly changing magnetic field of the Earth, whose recent history is known from studies of magnetized rocks. When the smooth variation due to this is removed, the carbon data still show a number of striking events. These could be misleading or spurious events in imperfect data, or they may be a record of variations in solar activity. We shall cautiously assume that they are, in fact, due to a changing Sun.

John Eddy has identified eighteen occasions in the past when the Sun was either quieter or more angry than usual, and has constructed a graph showing the Sun's behaviour right back to 5500 BC. A particularly noticeable feature of this curve is that we have been experiencing the crest of a new wave of solar activity in the last century or so. In the longer run of thousands of years the present level of solar activity, in terms of sunspots, extent of the corona, aurorae, and flares, may be unusually intense. Also in the data there is a vague hint, no more, that there is a cycle of solar activity with a period of about 2,500 years.

Now let us ask ourselves, are the changes in solar activity, as registered by trees, related to changes in our climate? Starting with historical data first, it certainly is rather intriguing that the last two minima of solar activity, occurring in the fifteenth and seventeenth centuries, coincide with prolonged and intensely cold weather. This protracted period of bitter weather is often called the Little Ice Age. Certainly this period is uniquely cold in the record of the past 3,000 years at least, coinciding with an unusually quiet Sun. In the seventeenth century the Baltic Sea froze over in winter. For many years

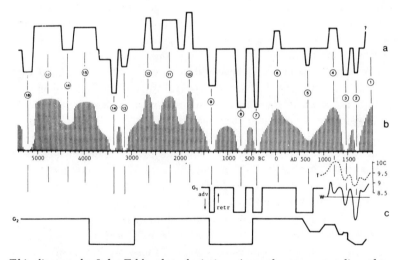

This diagram by John Eddy plots deviations from the average radiocarbon abundance in curve a. *Curve* b *shows the possible way in which the Sun may have varied to produce the effect in* a. *Curve* c *is in four parts:* G_1 *shows the advance and retreat of alpine glaciers, and* G_2 *displays the same information for a worldwide selection; curve* T *is an estimate of the average annual temperature in England and curve* W *a measure of the severity of the winter in Paris and London. Notice how the several curves clearly show the Maunder minimum, feature 2, the Spörer minimum, 3, and the medieval minor minimum at 5. (John A. Eddy)*

European rivers froze hard, and fairs were held on the river Thames at London. Population growth halted as winter's grip took its toll on agriculture. During the Little Ice Age the average annual temperature in England (for which country the best data exists) was about 1 degree centigrade colder than normal. Notice that a surprisingly small change has a large effect if it is sustained for decades or centuries at a time. Throughout the fifteenth and seventeenth centuries the northern hemisphere as a whole suffered an unremitting temperature reduction of 0.5–1 degree centigrade.

To go back in time more than a few hundred years, climatologists have taken a look at the recent fluctuations in the extent of glaciers, both in the European Alps and on a worldwide basis. When information on the advance of glaciers is compared to the Sun's radiocarbon signature, a good match is made. It has been found that when solar activity declines the ice advances, and when the Sun gets a little angry (as it is right now) the glaciers shrink back up the mountainsides. The glacial record also shows traces of the possible 2,500-year cycle that may be present in solar activity. The correlation is superficially impressive, but requires further data for confirmation.

Cautious researchers, such as Eddy, have rightly pointed out that we may be drawing the wrong conclusion. If fluctuations in the climate are somehow controlling the amount of carbon-14 that floats around in the atmosphere then a correlation of a climate indicator (such as glacial extent) with carbon-14 is bound to occur no matter what the cause. This is why the historical evidence on the Maunder minimum is so vital because we know certainly that no spots coincided with a cold climate. Similarly, the sparser data from the oriental sources give further support to the idea that a changing Sun shows up in the carbon record and the climatic fluctuations.

If we accept that the climatic changes over timescales of a century or a millennium are dictated by the Sun, there are still problems of interpretation. We would like to know how solar activity translates into a general warming of the Earth's climate. Perhaps this effect could be through magnetic influences, or it could be due to increases in the ultraviolet radiation from the Sun during periods of activity. The simplest mechanism of all is a change in the total flow of energy from the surface of the Sun. A change of around 1 per cent in the total flux would produce climatic variations on the scale of the Little Ice Age. An alteration of this extent would not have been detectable to astronomers if it were to have taken place gradually over the last hundred years, for example. And a drop in the central temperature of the Sun would explain away the neutrino problem.

Over the last few thousand years we have evidence that both the Sun and the climate have changed, and there does seem to be an observable connection between the two. For a long time astronomers and weather-watchers have toyed with a link between the 11-year solar cycle and the weather.

Our Sun — Our Future

2

Energy streams continuously from the hot Sun into the cold universe at an unimaginably high rate. The energy output of the Sun expressed in kilowatts is 3.8×10^{23} kW; for comparison we note that the energy consumption of most domestic appliances is 0.3–3 kW. Such large numbers are hard to grasp unless cast into more familiar language. An area of the solar surface no larger than a regular postage stamp (5 square centimetres) radiates about 30 kW. This is considerably more than the energy requirement of a small car. Yet another way of translating the solar energy is the true statement that in one second the Sun dumps more energy into the Universe than human civilization has generated in the whole of its history. Clearly the Sun is the most important source of energy for humans when we consider a sufficiently long time span.

The present state of planet Earth is that the world is controlled by a species that has developed technology, by which I mean the ability to make machines that consume energy and materials, in order to advance living standards above those that would prevail in an entirely natural (i.e., machineless) world. It is well known that the average living standard enjoyed by humans in various parts of the world is directly related to the amount of energy available cheaply to the local community. Here cheaply means of low cost relative to the cost of labour; so in rural India, for example, most forms of energy are very expensive apart from locally gathered dried cowdung and firewood, neither of which is suitable for driving machinery. On the other hand, Canada and the USA have the highest rates of energy consumption per head of population and in those countries the standard of living is very high when measured in purely material forms. Lifestyle then is currently strongly conditioned by a plentiful supply of energy that is cheap relative to labour and raw material costs; this ensures that homes can be comfortably heated or cooled and may be spacious if land is not expensive; transport too is cheap, and manufactured goods relatively inexpensive. Curiously, although both energy and raw materials are needed to manufacture goods, local availability of cheap energy is normally more important to an economy than local supplies of raw materials. For example, the United Kingdom has plentiful and diverse energy supplies but very few primary materials.

In his book *Ten Faces of the Universe* the astrophysicist and fiction writer Sir Fred Hoyle made a number of interesting observations on the energy consumption of the human species. For example, we consume about fifteen times as much energy inanimately (heating, fuel, machines) as we get from food. Technology means that, averaged over the globe, humans have fifteen times more energy at their disposal than a species that relies only on food. In fact the ratio of inanimate energy consumption to food consumption is a crude measure of the level of technology and the degree to which the planet's resources have been exploited. To give another example, in the Classical World, where the main source of energy for the wealthy classes was slave labour, the ratio of inanimate energy to food energy was probably less than one, and this small value accounts to a great extent for the difference in lifestyles between the civilizations of Greece and Rome and the industrial society of today. A civilization that could secure supplies of, say, a thousand times as much inanimate energy as food energy would either be extremely wasteful, or much further advanced than our own, or a combination of both of these.

Currently there is a good deal of distress in the highly developed societies, such as North America, Europe and Australasia, over the recurring energy crisis. This concern centres around supposedly diminishing supplies of oil, ever-rising prices of energy, and a consequent inability of freely elected (democratic) governments to contain inflation. Related to this is a feeling that resources in general are running out and that we are about to be buried by our own garbage, or choked by atmospheric pollution, or fried by radioactive waste. A good deal of this negative thinking is engendered by politicians, commentators and commercial interests concerned only with strictly short-term problems, such as next year's elections, tomorrow's newspapers or generating cash for the next round of pay rises. Viewed on a cosmic perspective, globally and across a greater span of time, there is ultimately no intrinsic shortage of energy, nor is there any fundamental reason why recycling of materials and the invention of synthetics should not supply sufficient material comforts. By the words 'ultimately', 'intrinsic' and 'fundamental' I mean that in the final analysis the resources of the planet are adequate for all when harnessed wisely and equitably, by mankind, in a manner that takes proper account of the total available global resources and the true needs of the human species generally. Of course, the key to this ideal situation is a correct appreciation of the roles that solar energy, nuclear energy and fossil energy have to play in a world that is planned along scientific principles and with a proper regard for the dignity and rights of all mankind. It is my personal opinion, influenced by Hoyle's conclusions, that society as it is presently organized is essentially geared around a disgraceful scramble to grab cheap energy today with no regard, apart from newspaper doomsday talk, for the condition of the world in another fifty years. Apparently some economists discount the future at 10 per cent annually: on this basis a gallon of fuel burned today is treated as if it were about 200 times as valuable as a gallon of fuel that someone could burn in fifty years' time. That's why the oil producers are pumping it out of the ground as fast as possible!

Our energy-consuming society is almost entirely supported by fossil fuels: oil, coal and gas, although nuclear and hydroelectric sources are increasingly important of course. The fossil fuels are solar energy, originally stored in organisms, that has accumulated over hundreds of millions of years. It took half a billion years (500 million years) to lay down the deposits of coal and oil. As is well known a good deal of this legacy has been burned in only a century.

At present, fossil fuels still represent an abundant source of energy that is available at only moderate cost and effort, at least in comparison to the alternatives and substitutes. Since western governments are strongly motivated to keep consumer prices low, indeed artificially low in the case of energy, there is not yet any strong push to switch over to non-fossil energy supplies.

How long will the proven reserves of coal and oil last? The present reserves could yield around 10^{23} joules (a modest 300-watt appliance such as a food mixer uses 10^6 joules in one hour), which is a sufficient supply for the present rate of use until the year AD 2500. This projection assumes no growth in annual rates of consumption, which is certainly not the case at present.

Now suppose that there is an enormous effort to conserve fuel: by insulating homes, using smaller cars, eliminating waste in industry, and so forth. Even with a halving of the rate of consumption the reserves could not be extended beyond AD 3000. It is important to bear in mind too that certain types of deposit, such as tar sands, oil shale and sparse coal deposits, can only be recovered by expending at the extraction stage much of the energy content of the deposit. We see then that in a future time span, comparable to the interval between the Renaissance and the present day, a fundamental readjustment of energy supplies and demand has to take place, since the fossil fuels otherwise inevitably become exhausted.

Therefore we may conclude that if a high-technology society with comfortable living and elegant leisure is deemed to be a desirable goal for all humans, then it is absolutely essential to get access to other energy supplies well before the fossil supplies are worked out. Conservation measures do not remove the need to substitute eventually. They are a sensible method of playing for time even though they merely postpone the date at which it becomes inevitable.

With a switch to nuclear-energy sources the problems of supply and reserves essentially become trivial. The reserves of thorium and uranium, needed as feedstocks in conventional nuclear fission reactors, are huge, perhaps a million times more than those of the fossil fuels. In addition to this there are fantastic reserves of the heavy form of hydrogen, deuterium, in the oceans. These are sufficient to keep an energy-intensive society going for many millions of years. The problem here, however, is one of technology.

The Sun readily turns hydrogen into helium, but this has only been accomplished in rather small-scale experiments on Earth. Certainly no hydrogen fusion experiment has yet yielded more energy than went into the apparatus. Containing the deuterium gas at a very hot temperature has so far proven impossible because it simply cannot be held in a container long enough for fusion to work. Many ingenious methods have been suggested for trapping the hot gas by magnetic fields. Although governments have spent heavily on plasma and fusion research a workable reactor has yet to be built. Nevertheless the 'Sun in the laboratory' may be getting nearer. Experiments in which exceedingly powerful lasers blast droplets of deuterium-rich water have been encouraging. Plasma-containment times are steadily increasing. We may be within striking distance of an experimental fusion reactor in the foreseeable future.

Among the general public there is genuine fear of nuclear power, which is widely envisaged as being extremely dangerous. The track record of the reactors is, however, good. The most serious accident to date, that at Three Mile Island in 1979, was brought under control before a disaster of great proportions actually took place. The reactor probably came within fifteen minutes of catastrophic melt down and the plant was seriously damaged. Occasional scares of that type may, in fact, be beneficial to the extent that they keep the plant operators more vigilant and safety conscious. Nuclear power plants have as much safety as possible designed into them. Accidents are more likely as a result of a succession of human failings rather than isolated cases of human error. This also must be, and is, taken into account at the design stage.

I think it is legitimate to form a preliminary judgement of the nuclear debate on the basis of the evidence so far. Nuclear energy is now a significant source of electrical energy in Europe and the USA, although it currently furnishes only a small percentage of the total requirements of those regions. It has supplied this energy for a number of years without any catastrophic explosions, without any dramatic number of deaths among plant operators, and without subjecting the local population to sustained levels of radiation that are clearly harmful. Sufficient power stations are now operating for the occasional scare story to reach the newspapers, and there has been at least one really serious incident, that at Three Mile Island. There have also been deaths directly attributable to the radiation from power stations. The numbers are very small and have only involved site workers.

The arguments against power stations should be listened to and considered, but they do not seem to add up to a compelling argument when considered against the strictly limited potential energy of fossil fuels. The problem of nuclear waste is an issue that the industry has got to face fair and square. A technology far ahead of our own in terms of energy usage would (by definition) have to have developed foolproof methods of dealing with radioactive wasteheaps. At the present level of waste production the enormous capacity of the oceans to absorb and diffuse toxic substances should not be overlooked, for their total volume already holds far more radioactivity than would be contributed by power-station waste.

It has seemed to me that the anti-nuclear lobby frequently disregards the heavy cost in human suffering and death that the conventional fuels involve: miners trapped or killed or prematurely retired through lung disease, divers killed in offshore oil-rig operations, employees killed in refinery accidents, explosions of fuel tankers at sea and on the road. Worldwide the death toll must run into thousands annually. Add to this the pollution of the atmosphere and the deaths that this pollution must cause, particularly among those who are already at risk through respiratory diseases. The record of fossil fuel is not so clean, is it?

The nuclear debate balances the theoretical worst case accident against the fossil fuel accidents that society blandly accepts, without feeding in the risk factors. The risks are very tiny (but still not zero) for nuclear accidents and large for employees in fossil fuel extraction and refinement. What has to be decided is whether we want western civilization to continue in more or less its present form—in which case nuclear programmes have to be accelerated, and the human species has to come to terms with the risks—or whether society is to switch to a lifestyle that consumes far less energy, probably has far less material goods and medical care, and therefore a lower life expectancy anyway. My guess is that most people, backed up against a wall, would prefer to live in the nuclear shadow rather than quaint rural poverty.

It is certainly desirable to proceed cautiously with any nuclear programme; fortunately there is a way of postponing hard decisions and perhaps managing with a limited dependence on nuclear energy. Time can be gained by much wiser use of the energy streaming in, free of charge, from the Sun. Every square metre of the Earth's surface receives over a kilowatt of energy when the Sun is directly overhead. The amounts are somewhat smaller for the real case with the Sun striking the ground or a glazed surface at an angle, but nevertheless the amount is of the order of one kilowatt per square metre. The roof area of even a small house collects about 1,000 kilowatt-hours (equivalent to 1,000 units of domestic electrical energy) in a single sunny day. For my own house, three sunny days in the summer months result in a larger energy input than our entire annual use of electricity for lighting and appliances. After a further ten days of sunny skies the roof has absorbed the same energy as the boiler consumes in the whole winter heating season. So we see that the potential value of solar energy is very high as a replacement for some of the present energy sources.

Two applications of solar energy need to be distinguished. These are solar-heating schemes and the generation of electricity directly from sunlight. The basic principles of solar heating are simple enough: trap and store the Sun's heat when the Sun is shining and use it later for heating. There are many ways of implementing this principle. To explore them all would take another book, but I will mention some of them briefly.

In the home the energy cost of heating the domestic hot water can be reduced if the water that enters the main tank is heated first, so that gas or electricity are needed only for supplementary heating. A convenient way of doing this preliminary heating is to pump water in a continuous cycle through a black radiator panel fixed on the roof and then through to the storage tank that is used to top up the hot-water tank. In England a simple system like this will provide adequate hot water almost free of charge in the summer. Indeed electronic controls are needed to prevent everything overheating on a really hot day. For the rest of the year, except perhaps midwinter, some useful heat is generated.

At present it is very expensive to heat an entire house by the Sun alone because solar-energy houses are still experimental. In a nutshell, expensive experimental devices are competing with underpriced energy. This situation will change gradually as the cost of fossil fuels rises in real terms. An encouraging sign in Europe and North America is the much greater awareness on the part of architects, designers and those who use buildings of the benefits of thick insulation and careful positioning of windows to admit the winter Sun. A house in which the main living rooms face south will be cheaper to heat in winter than one in which the windows face north. In hot climates, such as the southern USA and northern Australia, the opposite considerations apply: put the windows on the north side (USA) and south side (Australia) to reduce the cooling costs in the summer. Experimental houses have already been constructed in many countries in order to gain new insights into what can be accomplished. Right now such a house probably costs two or three times as much as a conventional house. If the cost differential came down to 10 or 20 per cent extra such houses would be a good buy. There is also great potential in countless commercial and industrial applications for a more sensible understanding of the Sun and its effects on buildings. Tinted glass, for example, is now standard in large office blocks partly because it reduces the cooling costs in sunny weather. There are leisure applications of solar heating too. Outdoor swimming pools can be covered with an insulating plastic blanket in contact with the water. This is particularly effective in reducing evaporation, the major cause of cooling, and therefore in raising the pool temperature.

So far I have considered some possibilities on the domestic scale. It is also going to be important, in desert areas where permanent sunshine is the norm, to generate electricity from solar energy. If this can be made to work reasonably cheaply it will diminish a major problem associated with using solar energy: you do not need much energy in deserts, where few people live, but you need a great deal in the major cities situated in cloudy temperate zones (New York, London, Moscow); this energy can be most readily supplied as electrical energy generated in the deserts. The total energy requirements of the United States, for example, can be met by solar collectors covering one-tenth of the state of Arizona and operating at only 10 per cent efficiency. Already a major experimental facility in New Mexico is generating energy by concentrating the Sun's rays onto a furnace. The high-pressure steam raised in the furnace can be used to drive a generator, just as happens in a fossil-fuelled power station.

At Albuquerque, New Mexico, the Sandia Laboratories has an experimental facility. This has been used to test the design of solar-powered electricity generating plants. The Sandia collector concentrates 5 megawatts of solar thermal energy on a receiver. A transfer medium, such as water, is circulated through the receiver where it is turned into high-pressure steam at nearly 1,000°C that can be used to drive a steam turbine generator. At Barstow, in California, a plant generating 10 megawatts is to be constructed, perhaps the first of several.

No collector on the Earth will be able to deliver solar energy at night. To get round this there are ambitious futuristic plans to build huge solar collectors that would be orbiting in space. These would possibly beam energy to the Earth in the form of microwaves. Ground stations would tune into this powerful electromagnetic radiation and convert it to conventional electrical energy. Preliminary studies suggest that such power stations, in Earth orbit, are not prohibitively expensive relative to the value of the energy they will yield. Construction would involve techniques already tried on much smaller scales in space.

As regards solar energy stations in space it is worth noting that they can, in principle, deliver an abundant supply of energy. But to reach

A solar receiver panel undergoing tests at the Sandia Laboratories test facility. The panel is on a 60-metre-high tower. Solar energy is concentrated by glass mirrors that track the Sun. The 222 mirrors are capable of directing up to 5 megawatts onto the receiver panel. (Sandia Laboratories, New Mexico)

that supply demands a vast investment and the use of advanced technology. This illustrates the general point that advanced technologies lead to greater supplies of energy as new techniques are invented and applied. There is no shortage of energy as such, just shortages of particular types of energy. In the longer term it would be folly if the easily accessible fossil fuels were run so low that the technology needed to construct nuclear plants or space stations simply could not be supplied by an energy-starved economy. The result of such folly would be a catastrophic and irreversible collapse of the human species. For this reason the energy sources that require advanced technology should be introduced or greatly expanded as soon as possible before it is too late.

The political unrest in several crucial supply areas in the 1970s has had an effect that could be beneficial in the long run. We are all more aware now of how dependent is the complex society of today on energy supplies. We are also aware that these supplies will not last for ever, and that in the short run they may be subject to politically motivated interference. Over a timescale that is considerably shorter than the span of recorded human history a fundamental change has to occur. Man must either learn to do without energy, and thus revert to a near-animal state, or he must learn to harness the two virtually limitless supplies: the energies within the nucleus and those from the daytime star, our Sun.

Bibliography

GIORGIO ABETTI, *The Sun*, Faber and Faber, 1957.

A classic account of solar physics and astronomy, first published in 1934, and accepted as the standard work up until the mid-1950s.

ANTHONY F. AVENI, *Archaeoastronomy in Pre-Columbian America*, University of Texas Press, 1975.

The first detailed treatment of primitive astronomy in the Americas to be published as a book. A fascinating guide to myth and folklore.

W. M. BAXTER, *The Sun and the Amateur Astronomer*, David and Charles, 1973.

A description of the observations that an amateur astronomer can undertake.

A. BRUZEK and C. J. DURRANT, *Illustrated Glossary for Solar and Solar-Terrestrial Physics*, D. Reidel, 1977.

A research-level text which defines many of the terms used in solar physics. Has extensive bibliographic references.

JOHN A. EDDY, *The New Solar Physics*, Westview, 1978.

An advanced review of current problems in solar physics.

————, *A New Sun: The Solar Results from Skylab*, NASA, 1979.

A popular account of the Skylab mission and the information it gave us about the Sun.

JOHN GRIBBIN, *Climatic Change*, Cambridge University Press, 1978.

A definitive account, at an advanced level, of the causes and effects of our changing climate.

MICHAEL KENWARD, *Potential Energy*, Cambridge University Press, 1976.

The author looks at the future for energy research and development, including the applications of solar energy.

FRED HOYLE, *On Stonehenge*, W. H. Freeman, 1977.

Was Stonehenge a solar observatory and eclipse predictor, as this book suggests?

FRED HOYLE, *Ten Faces of the Universe*, Heinemann, 1977.

A provocative, profound and controversial book which presses the case in favour of nuclear and solar energy.

DAVID K. McDANIELS, *The Sun: Our Future Energy Source*, John Wiley, 1979.

A survey of the theory and application of solar energy and collectors.

SIMON MITTON, *The Crab Nebula,* Faber and Faber, 1979.
Gives a description of the later stages in the life of a star.

SIMON MITTON, *The Cambridge Encyclopaedia of Astronomy,* Jonathan Cape, 1978.
This comprehensive work includes chapters on the Sun and the evolution of stars.

H. K. MOFFATT, *Magnetic Field Generation in Electrically Conducting Fluids,* Cambridge University Press, 1978.
An advanced account of dynamo theory.

JAMES MUIRDEN, *Astronomy with Binoculars,* Faber and Faber, 1976.
Includes a chapter on solar observations with binoculars. A useful guide to observing using limited equipment.

THOMAS RACKHAM, *Astronomical Photography at the Telescope,* Faber and Faber, 1972.
Includes a section on solar photography.

Index